デザインに即戦力
Photoshop厳選ブラシ集

Rutles

Adobe、Adobe Photoshop、Adobe Illustrator は、Adobe Systems Incorporated（アドビシステムズ社）の商標です。
Macintosh は Apple Computer,Inc. の各国での商標もしくは登録商標です。
Windows は米国 Microsoft Corporation の米国およびその他の国における商標または登録商標です。
その他本書に記載されている会社名、製品名は、各社の登録商標または商標です。

本書内容については、間違いがないよう最善の努力を払って検証していますが、著者および発行者は、
本書の利用によって生じたいかなる障害に対してもその責を負いませんので、あらかじめご了承ください。

はじめに

本書は主に海外のアーティストやデザイナーが作成した、Photoshop用のブラシプリセットの中からデザイン性に優れ、利用価値の高い215プリセットを選び、紹介した素材集です。

本書で紹介しているブラシは、Mac、Windowsその両方のPhotoshopで使うことができ、高解像度のものも含まれていますので、印刷用途からWeb作成まで幅広く利用できます。

紹介したブラシプリセットはテイスト、ジャンル毎に10のカテゴリに分け、それぞれのプリセットを使った作例を全部掲載していますので、目的のプリセットがすぐに探し出せ、またブラシのイメージも掴み易い構成になっています。

○ デザインのパーツが欲しいとき
○ 時間がなく、急いで作業しなければならないとき
○ デザインのクオリティをあげたいとき
○ デザインのアイデアが欲しいとき
などなど

本書のブラシプリセットは、このようなシーンで役に立つことと思います。

是非ブラシを使いこなして、素敵で楽しいデザインワークにお役立てください。

ラトルズ編集部

素材カタログの見方

Brush Number
053　**butterfly brushes set4**　Mohaafterdark

Ornamental Butterfly

📁 FLORAL » 📁 053-butterfly brushes set4

FLORAL

個人利用 ⃝　商用利用 ✕　※1

deviantART　http://mohaafterdark.blogspot.jp

権利に関して

素材の使用に関しては、ブラシカタログの下に 個人利用 ◯ 商用利用 ◯ で掲載しています。基本的に個人利用は、全て OK です。商用利用に関しては、ブラシにより違いがありますので、確認してください。特に条件などがある場合は、※、※ 1、※ 2、※ 3 で下記に補足をいれています。

― ブラシナンバー
― ブラシ名
― ブラシ作者名

FLORAL ≫ 053-butterfly brushes set4

deviantART　http://mohaafterdark.blogspot.jp

― ブラシを使った作例

いくつかのブラシでは、高解像度のものや、1 ファイルのみ利用できるものがあります。この場合、さらに他のファイルを使いたいときには、購入することで、利用することができるようになります。

販売している作者の例

http://env1ro.deviantart.com/
Gallery から Photoshop Brushes のカテゴリからブラシをクリックすると購入サイトへリンクしています。

※　商用利用は、クレジット希望。寄付歓迎。

※ 1　個人利用、商用利用とも◯
　　　個人利用、商用利用ともに利用可能です。改変は禁止。

― ブラシを収録したフォルダ
― 登録されているブラシ

※ 2　個人利用は◯、商用利用は×
　　　商用利用には、利用できません。
　　　作者によっては、メールで問合わせしてみましょう。

― ブラシ作者の Web ページの URL

個人利用、商用利用の可否を表示しています。また権利などの注釈も記載しています。

※ 3　Brusheezy ルール
　　　個人利用は◯、商用利用は作者からの直接の了解がない場合は「作者名 /brusheezy.com」のクレジットが必要。

― カテゴリ

Index

はじめに

本書の読み方 ・・・・・・・・・・・・・・・・・ 4

プリセットのインストール方法 ・・・・・・・・・・ 8

ブラシの使い方 ・・・・・・・・・・・・・・・ 10

素材のダウンロード方法 ・・・・・・・・・・・ 14

SPLATTER ・・・・・・ 15

PAINT ・・・・・・・・ 39

FLORAL ・・・・・・・ 57

NATURE ・・・・・・・ 99

ABSTRACT ・・・・ 129

EFFECTS · · · · · · 151

GRUNGE · · · · · 163

VECTOR · · · · · · 177

ORNAMENT · · 191

VARIETY · · · · · 209

プリセットのインストールと追加方法

本書に掲載している素材は、Photoshop 用のブラシプリセットをパソコンにインストールすることで使用できます。ブラシプリセットを直接ダブルクリックして読み込むこともできるので、好みに合わせて追加方法を選んでください。

【1】フォルダをコピーする

インストールしたいブラシのフォルダをデスクトップなどにコピーします。フォルダ内に拡張子「***.abr」が付いたファイルが、インストールに必要なブラシプリセットです。そのファイルを下記に示した所定のフォルダ内に入れます。削除する場合は、control キー＋クリックでメニューを開き、[ゴミ箱に入れる] を選びます。

```
Win版   コンピュータ C/ProgramFiles/Adobe Photoshop XXX/Presets/Brushes
Mac版   Macintosh HD/アプリケーション/Adobe Photoshop XXX/Presets/Brushes
```

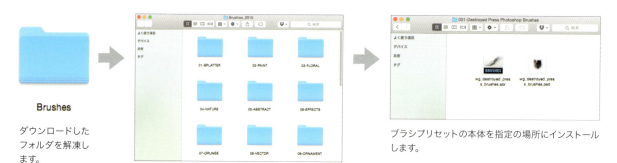

ダウンロードしたフォルダを解凍します。

一括ダウンロードした場合は、カテゴリごとにブラシが収録されています。

ブラシプリセットの本体を指定の場所にインストールします。

Mac の場合

[注] Photoshop でブラシプリセットを保存する場合、初期設定では、（ユーザー）/ ライブラリ /Application Support/Adobe/Adobe Photoshop/Presets/Brushes フォルダが選択（推奨）されています。このフォルダは、不可視フォルダ「ライブラリ」の階層にあるため、通常の操作では表示されません。表示する場合は、[移動] メニュー→[フォルダへ移動] を選択し、[フォルダの場所を入力] に「~/ ライブラリ」を入力して、[移動] をクリックするか、[option] キーを押しながら [移動] メニューをクリックするとライブラリフォルダがメニューに表示されます。

【2】プリセットを読み込む

Photoshop を起動します。[ブラシツール] を選択し、[ウィンドウ] メニュー→ [ブラシプリセット] で [ブラシプリセット] パネルを表示します。パネルメニューを表示すると、インストールしたブラシプリセットがリストアップされているので選択します。「現在のブラシを（***）のブラシで置き換えますか？」のアラートで [追加]、または [OK] をクリックします。

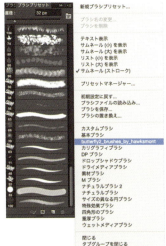

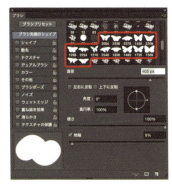

[注] プリセットの読み込みは、[ブラシツール]を選択して表示されるオプションバーでも行えます。[クリックでブラシプリセットピッカーを開く] をクリックして、歯車アイコンをクリックすると、[ブラシプリセット] パネルと同様のパネルメニューが表示されます。

■COLUMN

もっと簡単にブラシを読み込む方法！

コピーしたブラシプリセットファイルのアイコンをダブルクリックします。これだけで、[ブラシプリセット] パネルに追加されます。追加したブラシプリセットは、[プリセットマネージャー] で削除や名前変更などの管理ができ、複数のブラシプリセットを頻繁に入れ替えたい場合など、Photoshop への負担も軽減できます。何よりも簡単でスピーディ。デメリットは、Photoshop の設定を初期化したときなどに情報が失われてしまうことです。ブラシプリセットを自分で管理できる人にオススメ！

■COLUMN

プリセットマネージャから読み込む

Photoshop を起動します。[編集]メニュー→[プリセット]→[プリセットマネージャ]を表示します。右側の[読み込み]をクリックして、ブラシのプリセットを選択します。プリセットマネージャに戻って [完了] をクリックします。これで読み込むことができました。

[注] 本書で紹介しているブラシは、Mac/Windows で動作確認しておりますが、ブラシの制作時期が古いものは、環境によって動作しないものもあります。

ブラシパネルの基本操作

ブラシの機能拡張を含んだ詳細な設定ができるパネルです。ブラシの角度や真円率、間隔を組み合わせて設定できるほか、指定ステップによるフェード、サイズ、色、不透明度などをランダムに変化できます。

【1】［ブラシ］パネルを表示する

［ツール］パネルから［ブラシツール］を選びます。［ウィンドウ］メニューから［ブラシ］を選ぶ、または、オプションバーで［ブラシパネルの切り替え］をクリックします。

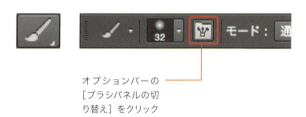

オプションバーの［ブラシパネルの切り替え］をクリック

【2】ブラシプリセットを選択する

［ブラシ］パネルで、［ブラシプリセット］から任意のブラシプリセットを選択します。

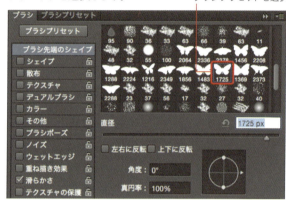

ブラシプリセットを選択

【3】ブラシの大きさを変える

ブラシプリセットサムネールの下に書かれてある数字は、元のブラシの大きさを示すもので、縦横の最大値をpixel単位で表示されています。

縦横の最大値が適用されている

オリジナルのサイズに戻す

スライダーを移動して、サイズを変更する

【4】ブラシの向きや角度を変える

［左右に反転］と［上下に反転］のチェックボックスで、元のブラシの向きを変えることができます。［角度］は、-180°～180°の数値が入力でき、初期設定は0°です。［真円率］は、角度0°に対する変形を百分率（パーセント）で入力します。［ブラシの角度と真円率を設定］では、［角度］と［真円率］をドラッグで設定することができます。

［角度］と［真円率］を設定

ブラシの形をランダムに変える

[ブラシ] パネルの [シェイプ] は、[ブラシ先端のシェイプ] で設定した [ブラシプリセット] のブラシ形状を、ストローク内で変化させる設定が行えます。

【1】シェイプを選択する

[シェイプ] をクリックして、チェックマークを入れます。

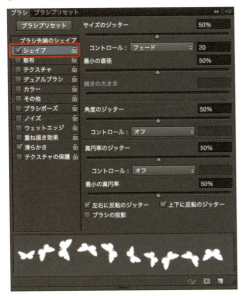

【2】ジッターを設定する

[サイズのジッター]、[角度のジッター]、[真円率のジッター] にそれぞれの変化率を設定します。

サイズのジッター
ブラシサイズの変化を設定します。[コントロール] に [フェード] を選択すると、指定したフェードステップ数で、初期値から [最小の直径] まで、ブラシサイズの変化を徐々に小さくしていくことができます。

角度のジッター
角度の変化を設定します。[コントロール] に [フェード] を選択すると、指定したフェードステップ数で、角度が 0°から 360°に変化します。

真円率のジッター
真円率の変化を設定します。[コントロール]に[フェード]を選択すると、指定したフェードステップ数で、真円率が 100%から [最小の真円率] まで、真円率の変化を徐々に変えていくことができます。

【3】ドキュメントをドラッグする

ブラシの [間隔] に「175%」を設定し、ドキュメントをワンストロークでドラッグします。すると、ブラシのサイズや角度を不規則に変化させることができます。

ワンストロークでドラッグする

■COLUMN

ジッターを 50%から試す！

ジッターとは、「不規則な揺れやゆがみ」の変化率です。ジッター 0%では変化がなく、100%で最大になります。しかし、これらの変化率は、ブラシの形状やサイズによって見え方が違うので、どこかつかみ所のないものです。設定に迷ったら中間値の 50%を試してみましょう。

ブラシを散りばめてペイントする

[ブラシ] パネルの [散布] は、[ブラシ先端のシェイプ] で設定した [ブラシプリセット] のブラシを、ストローク内で散りばめる設定が行えます。

【1】散布を選択する

[散布] をクリックして、チェックマークを入れます。

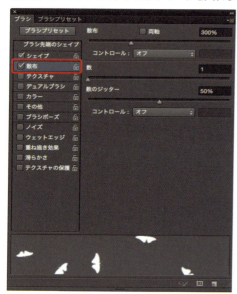

【2】分布方法とジッターを設定する

[散布]、[数]、[数のジッター] にそれぞれの散布率を設定します。

散布

ストローク範囲の分布方法と最大散布率を設定します。[両軸] オプションは放射状に分布、チェックマークを外すと垂直方向に分布します。[コントロール] に [フェード] を選択すると、指定したフェードステップ数で、最大散布率から「散布なし」に分布が変化します。

数

描点の個数（ブラシの密度）を設定します。数値が大きいほど密度が高くなります。

数のジッター

描点が分布する変化を設定します。[コントロール] に [フェード] を選択すると、指定したフェードステップ数で、[数] で設定した数値から「1」に分布が変化します。

【3】ドキュメントをドラッグする

[シェイプ] の設定を有効にしたまま、ドキュメントをワンストロークでドラッグします。すると、個々に回転や縮小したブラシが、[散布] などの設定により散りばめてペイントすることができます。

COLUMN

フェードを効果的に使おう！

[コントロール] に [フェード] を選択すると、指定したフェードステップ数で、効果の変化を徐々に弱めていくことができます。ストローク範囲で、散布に変化を付けたいときは、[フェード] を試してみましょう！

ブラシの色をランダムに変える

［ブラシ］パネルの［カラー］は、［ブラシ先端のシェイプ］で設定した［ブラシプリセット］のブラシの描画色を、ストローク内で変化させる設定が行えます。

【1】カラーを選択する

［カラー］をクリックして、チェックマークを入れます。

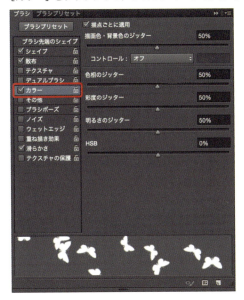

【2】ジッターを設定する

［描画色・背景色のジッター］、［色相のジッター］、［彩度のジッター］、［明るさのジッター］にそれぞれの変化率を設定します。

描画色・背景色のジッター
ペイント色の変化を設定します。［コントロール］に［フェード］を選択すると、指定したフェードステップ数で、現在の描画色から背景色に変化します。［描点ごとに適用］を有効にすると、ブラシの描点ごとに色が変化するので、同一のストローク内では縞状のさまざまなカラーが適用されます。

色相のジッター
色相の変化を設定します。［色相のジッター］に「0%」を設定すると、現在の描画色の色相がそのまま維持されるので、同系色の変化をつくり出すことができます。

彩度のジッター
彩度の変化を設定します。［彩度のジッター］に「0%」を設定すると、現在の描画色の彩度がそのまま維持されるので、彩度を統一した色の変化をつくり出すことができます。

明るさのジッター
明度の変化を設定します。［明るさのジッター］に「0%」を設定すると、現在の描画色の明度がそのまま維持されるので、明るさを統一した色の変化をつくり出すことができます。

【3】ドキュメントをドラッグする

［シェイプ］、［散布］の設定を有効にしたまま、ドキュメントをワンストロークでドラッグします。すると、ランダムに散布されたブラシが、［カラー］の設定によりさまざまな色に変化させてペイントすることができました。

■COLUMN

［描点ごとに適用］を効果的に使おう！

［描点ごとに適用］とは、カラーの変化をストローク内で行わないか、描点ごとに適用するかを切り替えるチェックボックスです。この項目が有効と無効では、大きな違いがあります。必ず現在の設定を確認しましょう。

収録素材について

本書で紹介しているブラシは、MacとWindowsのPhotoshopで使用することができます。一部の素材はMacのみの素材もありその場合は、作例の各ページに記載しています。基本的には、Photoshop CCで動作確認していますが、環境によっては動作しない場合がありますのでご了承ください。
素材は高解像度のものも含まれており、印刷をはじめとして、さまざまな用途に利用できます。

ブラシ素材は、作者の許諾を得て掲載しています。ブラシ素材は無料で配布していても著作権は作者本人にあります。そのため作者に無断での再配布や改変は禁じられています。カタログページには、個人利用、商用利用をわかりやすく記載しています。
商用利用がOKでも作者のホームページなどを訪問して確認するようにしましょう。

ブラシファイルは、カテゴリごとにzip形式で圧縮してあります。（一括ダウンロードも用意してあります）
ダウンロードには、ネットに接続できる環境が必要です。ダウンロード時間は、お使いの回線の速度によって大きくかわります。

データの構成

データは、10のカテゴリに分けてダウンロードすることができます。データは、圧縮されており解凍するには、パスワードが必要です。**パスワードは、240ページ奥付にある本書のISBNコード数字部分のみ、逆にした7ではじまる13桁の半角数字になります。**（一括ダウンロードもご用意していますが、ファイルサイズが約4GB弱ほどありますので、光など高速な通信環境でのダウンロードをお勧めいたします。）

データのダウンロード先

http://www.rutles.net/download/432/index.html

Brush Number
001
Destroyed Press Photoshop Brushes — Nathan Brown

📁 SPLATTER » 📁 001-Destroyed Press Photoshop Brushes

SPLATTER

 個人利用   商用利用

GraphicMonkee   http://www.graphicmonkee.com

Nathan Brown **Garbage Stain Brushes**

Brush Number **002**

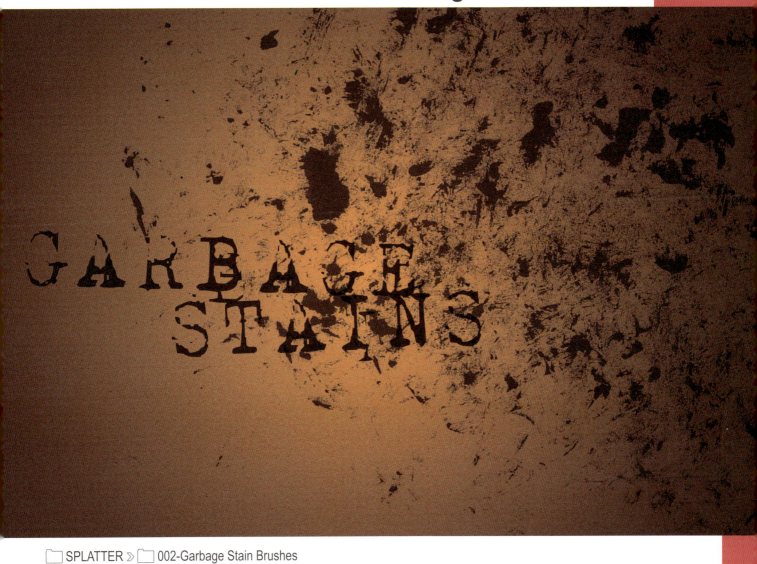

📁 SPLATTER » 📁 002-Garbage Stain Brushes

GraphicMonkee   http://www.graphicmonkee.com

SPLATTER

003 WINDBLOWN WATERCOLOR BRUSH SET
Brush Number — Nathan Brown

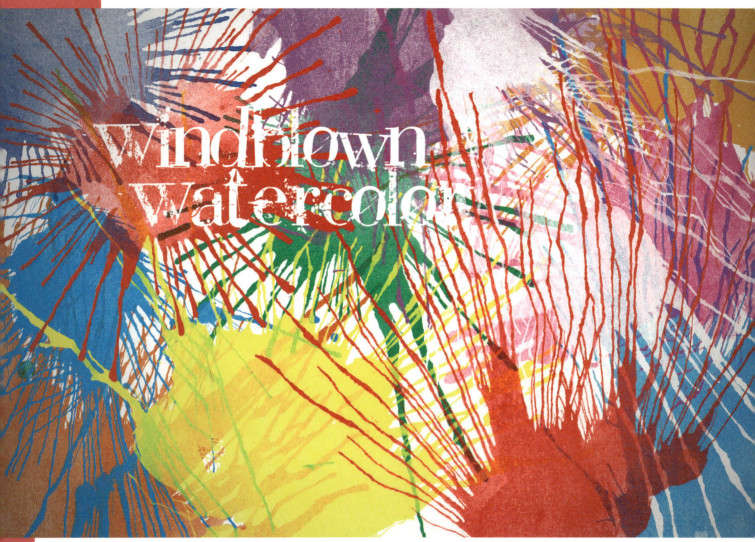

📁 SPLATTER » 📁 003-WINDBLOWN WATERCOLOR BRUSH SET

SPLATTER

個人利用 　商用利用

GraphicMonkee　http://www.graphicmonkee.com

Sylvain Bilodeau **15 Free Splatter Brushes** Brush Number **004**

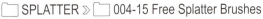
SPLATTER » 004-15 Free Splatter Brushes

Sylvain Bilodeau Digital Art　http://www.sylver.biz/

SPLATTER

Brush Number
005
12 FREE HQ Pro Splatter Brushes — Eldar Zakirov

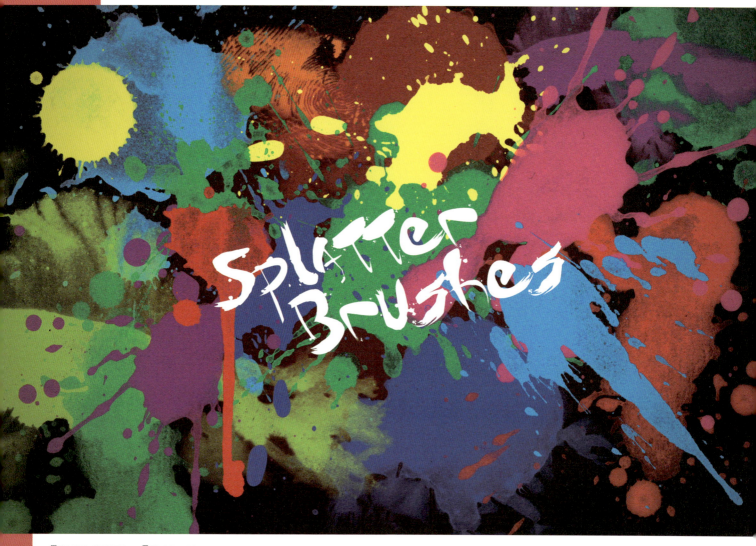

SPLATTER » 005-12 FREE HQ Pro Splatter Brushes

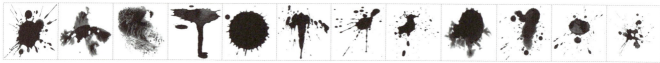

SPLATTER

 個人利用 　 商用利用

EldarZakirov　http://eldarzakirov.com/

Eldar Zakirov **CLAXA** Brush Number **006**

SPLATTER » 006-CLAXA

SPLATTER

 個人利用   商用利用

EldarZakirov   http://eldarzakirov.com/

Brush Number
007
spray-paint-photoshop-brush-Creative_Nerds Timothy Blake (Creative Nerds)

📁 SPLATTER » 📁 007-spray-paint-photoshop-brush-Creative_Nerds

SPLATTER

個人利用 　　商用利用

creative NERDS　http://creativenerds.co.uk

Javier Larios Barbosa **Vector Splatter**

Brush Number **008**

☐ SPLATTER ≫ ☐ 008-Vector Splatter

deviantART　http://javierzhx.deviantart.com

SPLATTER

Brush Number
009
Splatters Brushes FackFebruary (Jobey)

📁 SPLATTER » 📁 009-Splatters Brushes

SPLATTER

個人利用   商用利用

deviantART http://fackfebruary.deviantart.com

Phoenix **10 Extremely Large**

Brush Number **010**

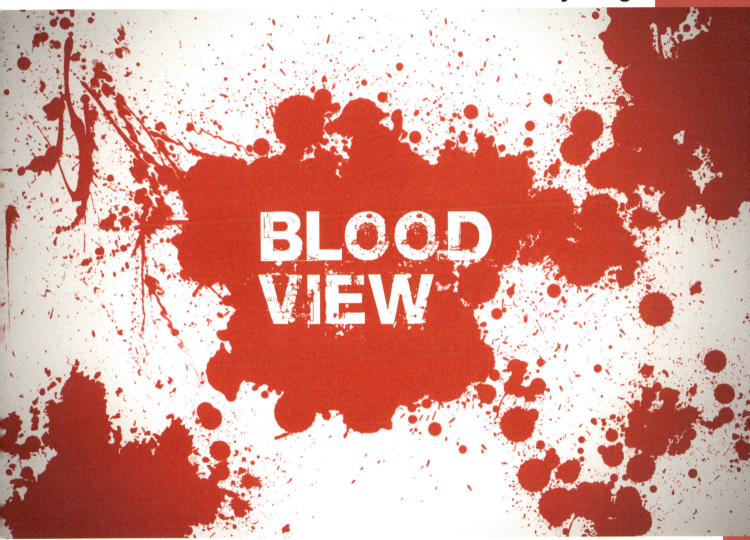

📁 SPLATTER » 📁 010-10 Extremely Large

The Forgotten Lair http://www.theforgottenlair.net

SPLATTER

Brush Number
011
Splatters hawksmont

📁 SPLATTER » 📁 011-Splatters

SPLATTER

個人利用 　　商用利用 　　　　　deviantART　http://hawksmont.deviantart.com

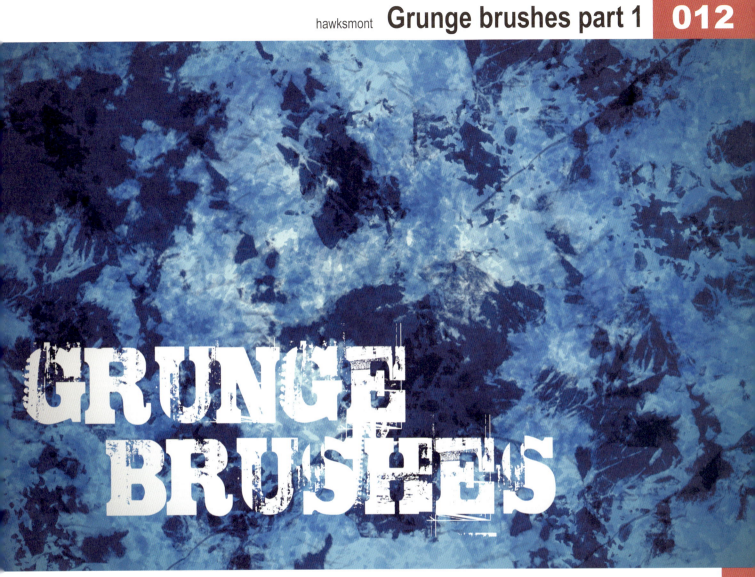

hawksmont Grunge brushes part 1

Brush Number **012**

📁 SPLATTER » 📁 012-Grunge brushes part 1

deviantART http://hawksmont.deviantart.com

SPLATTER

Brush Number
013
Splatter and Swirls Brushes Coby17 (Brenda Rivera)

📁 SPLATTER » 📁 013-Splatter and Swirls Brushes

SPLATTER

deviantART http://coby17.deviantart.com

jeff_finley **Go Media Spills & Splatters**

Brush Number **014**

📁 SPLATTER » 📁 014-Go Media Spills & Splatters

Brusheezy http://www.brusheezy.com/members/jeff_finley

SPLATTER

Brush Number 015
Watercolor Splatter Brushes — Geno Arguelles

SPLATTER » 015-Watercolor Splatter Brushes

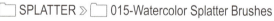

Geno Arguelles   http://genoarguelles.com

Brush Number 016

Geno Arguelles **Watercolor Strokes Paint Brushes**

📁 SPLATTER » 📁 016-Watercolor Strokes Paint Brushes

個人利用 　商用利用

Geno Arguelles　http://genoarguelles.com

SPLATTER

Brush Number 017
16 Splatter Brushes Cary_HMS

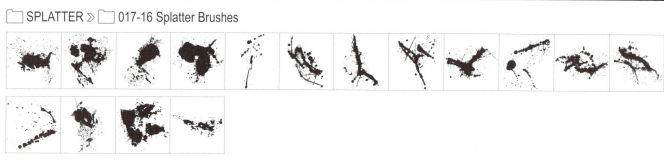

SPLATTER » 017-16 Splatter Brushes

個人利用 ○　商用利用 ○ ※3

Brusheezy　http://www.brusheezy.com/members/cary_hms

Brush Number
019 Ink Drop - 4 Brushes Jon Bee

 SPLATTER » 019-Ink Drop - 4 Brushes

SPLATTER

 個人利用   商用利用

WaterColor Reloaded

Przemyslaw 'env1ro' Szczepanski

Brush Number **020**

📁 SPLATTER » 📁 020-WaterColor Reloaded

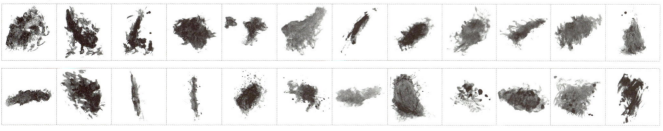

個人利用 ⭕   商用利用 ⭕   *

deviantART   http://env1ro.deviantart.com/

SPLATTER

Brush Number
021 Got Sprayed Photoshop brushes
Przemyslaw 'env1ro' Szczepanski

SPLATTER » 021-Got Sprayed Photoshop brushes

SPLATTER

個人利用 ⃝　商用利用 ⃝　*

deviantART　http://env1ro.deviantart.com/

Paint Lines brushes

Przemyslaw 'env1ro' Szczepanski

Brush Number **022**

SPLATTER » 022-Paint Lines brushes

 個人利用 〇　 商用利用 〇　*

deviantART　http://env1ro.deviantart.com/

SPLATTER

Brush Number
023 WaterColor EXTREMUM
Przemyslaw 'env1ro' Szczepanski

SPLATTER » 023-WaterColor EXTREMUM

個人利用 ◯ 商用利用 ◯ *

deviantART   http://env1ro.deviantart.com/

PAINT

Brush Number
024
Thick Paint Acrylic Brush Timothy Blake (Creative Nerds)

📁 PAINT » 📁 024-Thick Paint Acrylic Free Photoshop Brush Set

PAINT

個人利用 ◯   商用利用 ◯

creative NERDS   http://creativenerds.co.uk

Timothy Blake (Creative Nerds) **Make-up-smudges-Creative_Nerds**

Brush Number
025

📁 PAINT ≫ 📁 025-Make-up-smudges-Creative_Nerds

個人利用 ◯   商用利用 ◯

creative NERDS   http://creativenerds.co.uk

PAINT

Brush Number **026** pencil-strokes-brush-set-Creative-Nerds Timothy Blake (Creative Nerds)

📁 PAINT ≫ 📁 026-pencil-strokes-brush-set-Creative-Nerds

個人利用   商用利用

creative NERDS   http://creativenerds.co.uk

Brush Number
027

zummerfish **Zummerfish's Artistic N Texture Brushes**

PAINT » 027-Zummerfish's Artistic N Texture Brushes

deviantART   http://www.zummerfish.deviantart.com

Brush Number
028 Zummerfish's Acrylic Brushes for Photoshop zummerfish

ACRYLIC BRUSHES FOR PHOTOSHOP

📁 PAINT » 📁 028-Zummerfish's Acrylic Brushes for Photoshop

PAINT

個人利用 ⬜　　商用利用 ⬜ ※

deviantART　http://www.zummerfish.deviantart.com

Zummerfish's Blending Brushes

Brush Number 029

zummerfish

PAINT » 029-Zummerfish's Blending Brushes

個人利用 　商用利用

deviantART　http://www.zummerfish.deviantart.com

Brush Number
030

62 Ink Brushes FackFebruary (Jobey)

 PAINT » 030-62 Ink Brushes

個人利用 ◯   商用利用 ◯

deviantART http://fackfebruary.deviantart.com

FackFebruary (Jobey) **Pencil Brushes**

Brush Number **031**

📁 PAINT » 📁 031-Pencil Brushes

個人利用 ◯ 商用利用 ◯

deviantART   http://fackfebruary.deviantart.com/

PAINT

Brush Number
032 Danger Pig Brush Strokes Pack 01 dangerpig

PAINT » 032-Danger Pig Brush Strokes Pack 01

PAINT

 個人利用 ○   商用利用 × 要連絡

Brusheezy   http://www.brusheezy.com/members/vectoroom

Brush Number
AlicesPalette **Paint brush work** 033

📁 PAINT » 📁 033-Paint brush work

PAINT

個人利用 　　商用利用 　　　　　Brusheezy　http://www.brusheezy.com/brushes/2394-paint-brush-work

Brush Number
034
Splash 1.0 Robert Frank (vectoroom)

📁 PAINT » 📁 034-Splash 1.0

PAINT

個人利用 ○　　商用利用 ○ ※3

Brusheezy　http://www.brusheezy.com/members/vectoroom

James Myers (nineteeneightysevendesign) **Splatter Brush Set 0**

Brush Number **035**

📁 PAINT » 📁 035-Splatter Brush Set 0

個人利用 ◯ ※2　商用利用 ✕　　Brusheezy　http://www.brusheezy.com/members/nineteeneightysevendesign

Brush Number
036 Go Media Spray Paint jeff_finley

📁 PAINT ≫ 📁 036-Go Media Spray Paint

PAINT

個人利用 ◯   商用利用 ◯ ※3   Brusheezy   http://www.brusheezy.com/members/jeff_finley

Brush Number **037**

Javid Kazmi **Finger Paints Brush set**

PAINT » 037-Finger Paints Brush set

個人利用 ◯　商用利用 ◯

BrushKing   http://www.brushking.eu/307/finger-paints-brush-set.html

PAINT

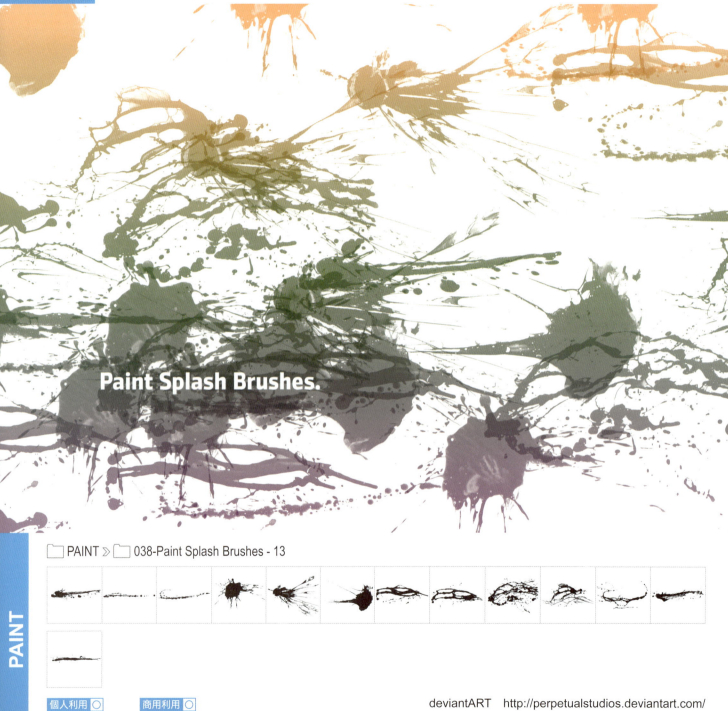

Timothy Blake (Creative Nerds) **Paint Borders**

Brush Number **039**

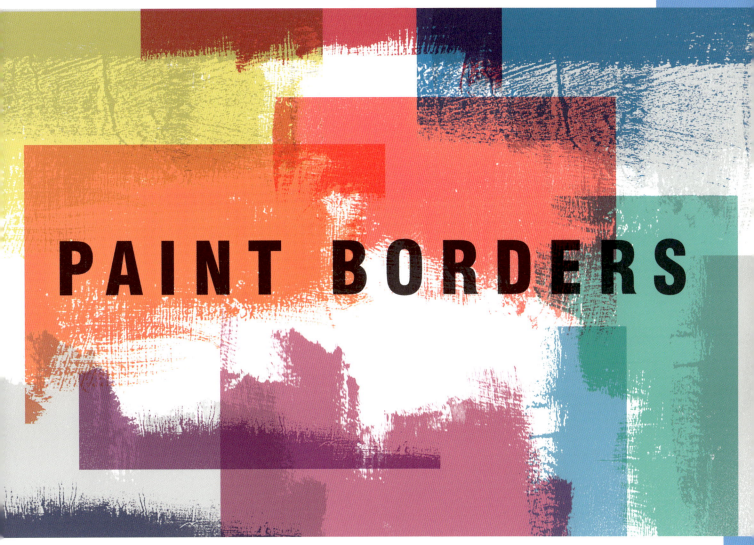

📁 PAINT » 📁 039-Paint Borders

個人利用 ○   商用利用 ○ *

creative NERDS   http://creativenerds.co.uk

Brush Number
040   Abstract Paint   Przemyslaw 'env1ro' Szczepanski

PAINT » 040-Abstract Paint

個人利用 ○   商用利用 □ *

deviantART   http://env1ro.deviantart.com/

Brush Number
041 Floral Ornaments Mohaafterdark

FLORAL » 041-Floral Ornaments

FLORAL

個人利用 ○　　商用利用 ×

POLICE Media　http://mohaafterdark.blogspot.jp

Mohaafterdark **Forever Hearts**

Brush Number **042**

FLORAL » 042-Forever Hearts

個人利用 ○ 商用利用 ✗

POLICE Media http://mohaafterdark.blogspot.jp

FLORAL

Brush Number
043

Hearts and Butterflies 1 Mohaafterdark

📁 FLORAL ≫ 📁 043-Hearts and Butterflies 1

FLORAL

個人利用 ⭕   商用利用 ❌

POLICE Media   http://mohaafterdark.blogspot.jp

Mohaafterdark **Hearts and Butterflies 2**

Brush Number **044**

FLORAL » 044-Hearts and Butterflies 2

POLICE Media   http://mohaafterdark.blogspot.jp

Brush Number
045
Hearts and Butterflies 3 — Mohaafterdark

FLORAL » 045-Hearts and Butterflies 3

FLORAL

個人利用   商用利用

POLICE Media http://mohaafterdark.blogspot.jp

Mohaafterdark **Ornament** — Brush Number **046**

FLORAL ≫ 046-Ornament

POLICE Media   http://mohaafterdark.blogspot.jp

FLORAL

Brush Number
047 Heart brushes set 2 Mohaafterdark

📁 FLORAL » 📁 047-Heart brushes set 2

個人利用 ⭕   商用利用 ❌

POLICE Media   http://mohaafterdark.blogspot.jp

Mohaafterdark **Heart brushes set 4**

Brush Number **048**

 FLORAL » 048-Heart brushes set 4

POLICE Media http://mohaafterdark.blogspot.jp

FLORAL

Brush Number
049
Heart brushes set 6 Mohaafterdark

 FLORAL » 049-Heart brushes set 6

FLORAL

POLICE Media   http://mohaafterdark.blogspot.jp

Mohaafterdark **Ornamental ButterFly**

Brush Number **050**

📁 FLORAL » 📁 050-Ornamental ButterFly

POLICE Media   http://mohaafterdark.blogspot.jp

Brush Number
051 Sticker with butterflies Mohaafterdark

📁 FLORAL » 📁 051-Sticker with butterflies

FLORAL

POLICE Media　http://mohaafterdark.blogspot.jp

butterfly brushes set1

Mohaafterdark

Brush Number **052**

Butterfly Brushes

FLORAL » 052-butterfly brushes set1

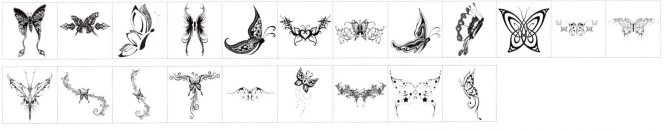

POLICE Media   http://mohaafterdark.blogspot.jp

Brush Number
053 butterfly brushes set4 Mohaafterdark

FLORAL » 053-butterfly brushes set4

個人利用 ○ 商用利用 ×

POLICE Media http://mohaafterdark.blogspot.jp

Mohaafterdark **Flowers OR** — Brush Number **054**

FLOWERS ORNAMENT

FLORAL » 054-Flowers OR

個人利用 ◯　商用利用 ✕

POLICE Media　http://mohaafterdark.blogspot.jp

FLORAL

Brush Number
055
Flowers and Swirls 2 — Mohaafterdark

📁 FLORAL ≫ 📁 055-Flowers and Swirls 2

FLORAL

個人利用 　商用利用

POLICE Media　http://mohaafterdark.blogspot.jp

Mohaafterdark **Flowers Rays** — Brush Number **056**

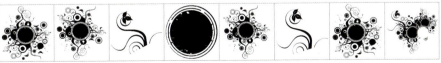

POLICE Media   http://mohaafterdark.blogspot.jp

Brush Number
057
Floral 19 Mohaafterdark

📁 FLORAL 〉 📁 057-Floral 19

個人利用 ◯   商用利用 ✕   POLICE Media   http://mohaafterdark.blogspot.jp

Mohaafterdark **Flowers Ornaments**

Brush Number **058**

📁 FLORAL » 📁 058-Flowers Ornaments

POLICE Media   http://mohaafterdark.blogspot.jp

FLORAL

Brush Number
059 Flowers Corners Mohaafterdark

📁 FLORAL ≫ 📁 059-Flowers Corners

FLORAL

個人利用 ⭕　商用利用 ❌　　　POLICE Media　http://mohaafterdark.blogspot.jp

Mohaafterdark flowers brushes set1

Brush Number 060

FLORAL » 060-flowers brushes set1

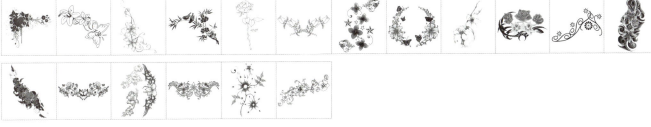

個人利用 ○   商用利用 ✕

POLICE Media   http://mohaafterdark.blogspot.jp

FLORAL

061 flowers brushes set 2 — Mohaafterdark

□ FLORAL » □ 061-flowers brushes set 2

POLICE Media   http://mohaafterdark.blogspot.jp

Mohaafterdark **ASdal Digital** — Brush Number 062

📁 FLORAL » 📁 062-ASdal Digital

POLICE Media   http://mohaafterdark.blogspot.jp

Brush Number
063
Floral 17 Mohaafterdark

📁 FLORAL ›› 📁 063-Floral 17

FLORAL

個人利用 ⭕ 商用利用 ❌   POLICE Media   http://mohaafterdark.blogspot.jp

Mohaafterdark **Floral 7** — Brush Number **064**

FLORAL » 064-Floral 7

 個人利用 ○   商用利用 ×   POLICE Media   http://mohaafterdark.blogspot.jp

Brush Number
065
Floral Circle 1 — Mohaafterdark

📁 FLORAL » 📁 065-Floral Circle 1

個人利用 ○   商用利用 ✕

POLICE Media   http://mohaafterdark.blogspot.jp

Mohaafterdark **Ornamental Butterflies 2**

Brush Number **066**

📁 FLORAL » 📁 066-Ornamental Butterflies 2

個人利用 ⭕ 商用利用 ❌

POLICE Media　http://mohaafterdark.blogspot.jp

FLORAL

Flowers Brushes — hawksmont
Brush Number 067

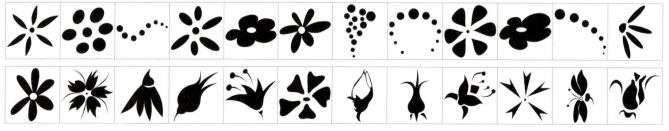

FLORAL › 067-Flowers Brushes

deviantART http://hawksmont.deviantart.com/

hawksmont **Bamboo Brushes**

Brush Number
068

📁 FLORAL ≫ 📁 068-Bamboo Brushes

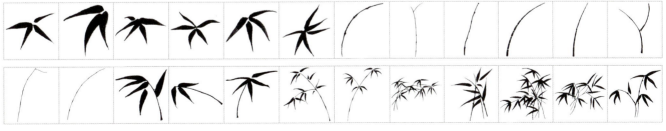

deviantART   http://hawksmont.deviantart.com/

069 Floral brushes hawksmont

📁 FLORAL » 📁 069-Floral brushes

deviantART http://hawksmont.deviantart.com/

hawksmont **Floral brushes II**

Brush Number **070**

Brush Number
071

Floral Elements Photoshop brushes Coby17 (Brenda Rivera)

📁 FLORAL ≫ 📁 071-Floral Elements Photoshop brushes

FLORAL

deviantART http://coby17.deviantart.com/

Coby17 (Brenda Rivera) **Beautiful Flowers Brushes**

Brush Number **072**

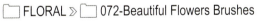
📁 FLORAL » 📁 072-Beautiful Flowers Brushes

deviantART http://coby17.deviantart.com/

FLORAL

Brush Number
073 Swirls Brushes Coby17 (Brenda Rivera)

 FLORAL » 073-Swirls Brushes

FLORAL

 個人利用 ○　 商用利用 ✕

deviantART　http://coby17.deviantart.com/

Coby17 (Brenda Rivera) **Cute Swirl Lines Brushes**

Brush Number **074**

📁 FLORAL » 📁 074-Cute Swirl Lines Brushes

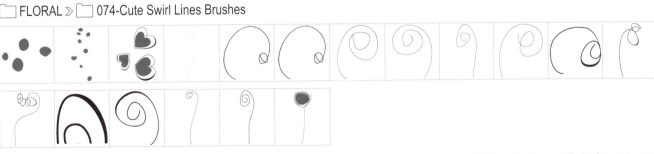

個人利用 　商用利用

deviantART　http://coby17.deviantart.com/

075 Floral swirls HD Photoshop Brushes — Coby17 (Brenda Rivera)

FLORAL » 075-Floral swirls HD Photoshop Brushes

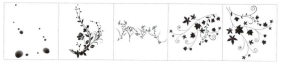

FLORAL

deviantART http://coby17.deviantart.com/

Coby17 (Brenda Rivera) **Flower Photoshop Brushes**

Brush Number **076**

FLOWER PHOTOSHOP BRUSHES

📁 FLORAL » 📁 076-Flower Photoshop Brushes

FLORAL

deviantART http://coby17.deviantart.com/

Brush Number
077 Sakura Flowers Brushes Coby17 (Brenda Rivera)

FLORAL » 077-Sakura Flowers Brushes

個人利用   商用利用

deviantART   http://coby17.deviantart.com/

Helenartathome **Nature1 Brushes**

Brush Number **078**

FLORAL » 078-Nature1 Brushes

個人利用   商用利用 ※1

Brusheezy　http://www.brusheezy.com/members/helenartathome

FLORAL

Brush Number
079

Photoshop Plant Brushes Helenartathome

📁 FLORAL » 📁 079-Photoshop Plant Brushes

FLORAL

個人利用 ⃝　商用利用 ⃝ ※1

Brusheezy　http://www.brusheezy.com/members/bsilvia

Przemyslaw 'env1ro' Szczepanski **Floral Abstract**

Brush Number
080

FLORAL ≫ 080-Floral Abstract

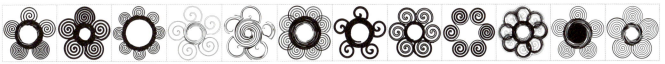

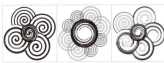

 個人利用 ○   商用利用 ○   *

deviantART   http://env1ro.deviantart.com/

FLORAL

Brush Number
081
FLORAL FANTASY Audee Mirza

FLORAL » 081-FLORAL FANTASY

FLORAL

 個人利用   商用利用

Audee Mirza | Logo & Web Designer   http://audeemirza.com/

Cloud Brushes

Brush Number 082 — Javier Larios Barbosa

NATURE » 082-Cloud Brushes

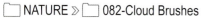

deviantART http://javierzhx.deviantart.com/

Delia Galhotra **Clouds pack 8 by Ailedda** Brush Number **083**

📁 NATURE » 📁 083-Clouds pack 8 by Ailedda

Delia Galhotra freelance phtographer http://deliagalhotra.wix.com/photography/

Brush Number
084
Photoshop Cloud Brushes With Commercial License Volume 1 Guy Dub

NATURE ≫ 084-Photoshop Cloud Brushes With Commercial License Volume 1

NATURE

Clouds Photoshop Brushes

Coby17 (Brenda Rivera)

Brush Number **085**

📁 NATURE » 📁 085-Clouds Photoshop Brushes

個人利用 ◯　商用利用 ✕

deviantART　http://coby17.deviantart.com/

Brush Number
086

5 Rain Brushes — Phoenix

Rhythm Of The Rain
If you want the rainbow, you gotta put up with the rain.

📁 NATURE » 📁 086-5 Rain Brushes

NATURE

個人利用 ◯   商用利用 ◯

The Forgotten Lair   http://www.theforgottenlair.net/

hawksmont **Snowflakes Brushes** Brush Number **087**

NATURE » 087-Snowflakes Brushes

個人利用 ◯   商用利用 ◯

hawksmont Universe :)   http://hawksmont.com/

NATURE

Brush Number
088 Drops Brushes Coby17 (Brenda Rivera)

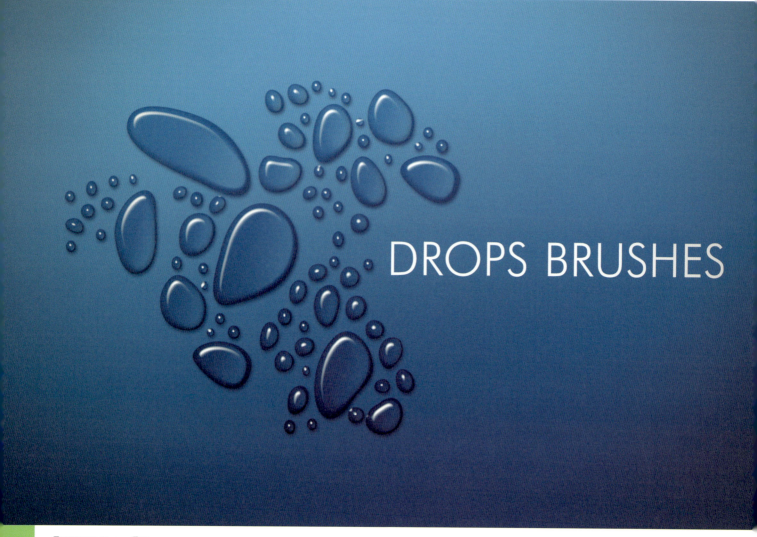

NATURE » 088-Drops Brushes

NATURE

個人利用 ○　　商用利用 ✗

deviantART　http://coby17.deviantart.com/

Lightning Brushes by Ailedda

Delia Galhotra — Brush Number 089

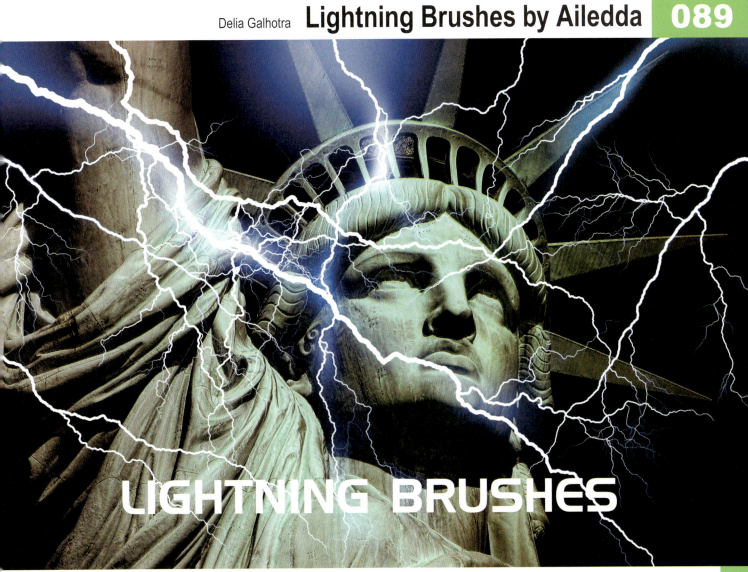

📁 NATURE » 📁 089-Lightning Brushes by Ailedda

Delia Galhotra freelance phtographer http://deliagalhotra.wix.com/photography/

Brush Number
090

25 Hi Res Smoke Brushes nadaimages

📁 NATURE ≫ 📁 090-25 Hi Res Smoke Brushes

NATURE

個人利用 ◯　　商用利用 ◯　　※3注　製品版には25個のブラシが用意されています。

Brusheezy　http://www.brusheezy.com/members/nadaimages

Jon Bee **Smoke Brushes - Six**

Brush Number
091

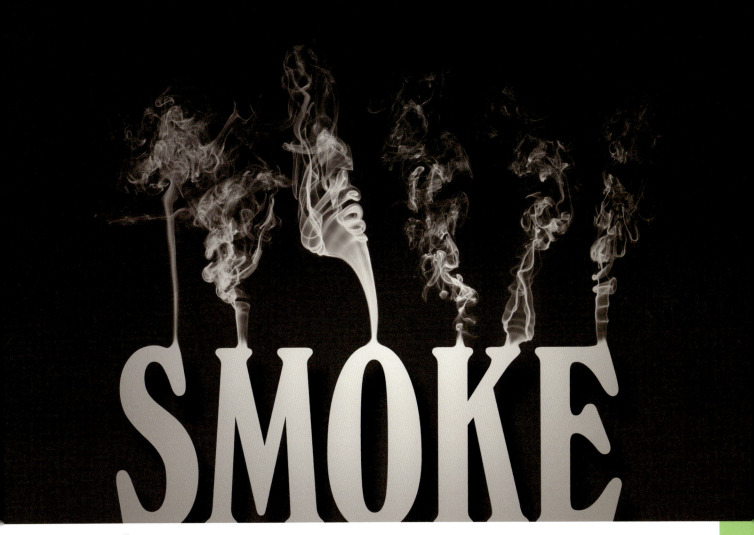

📁 NATURE » 📁 091-Smoke Brushes - Six

NATURE

Brush Number
092 Hi-Res Smoke PS Brushes Free Goodies for Designers

📁 NATURE » 📁 092-Hi-Res Smoke PS Brushes

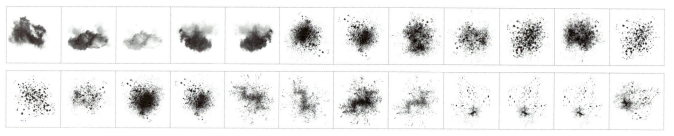

個人利用 ◯   商用利用 ◯   FREE GOODIES FOR DESIGNERS   http://freegoodiesfordesigners.blogspot.com/

Timothy Blake (Creative Nerds) **smoke-photoshop-brush-creative-nerds**

Brush Number
093

📁 NATURE ≫ 📁 093-smoke-photoshop-brush-creative-nerds

NATURE

個人利用 　　商用利用

creative NERDS　http://creativenerds.co.uk/

094 Zummerfish's Mystic Bubbles Brushes — zummerfish

NATURE » 094-Zummerfish's Mystic Bubbles Brushes

 個人利用 商用利用

deviantART http://www.zummerfish.deviantart.com/

Andrzej Walkowiak **Sparkling brushes**

Brush Number **095**

NATURE » 095-Sparkling brushes

deviantART http://lordandre.deviantart.com/

Brush Number
096 Soap bubbles Andrzej Walkowiak

📁 NATURE ≫ 📁 096-Soap bubbles

deviantART   http://lordandre.deviantart.com/

hawksmont **Bubbles**

Brush Number
097

📁 NATURE » 📁 097-Bubbles

個人利用 ◯　　商用利用 ◯

deviantART　http://hawksmont.com/

NATURE

Brush Number
098
Bubbles Brushes Coby17 (Brenda Rivera)

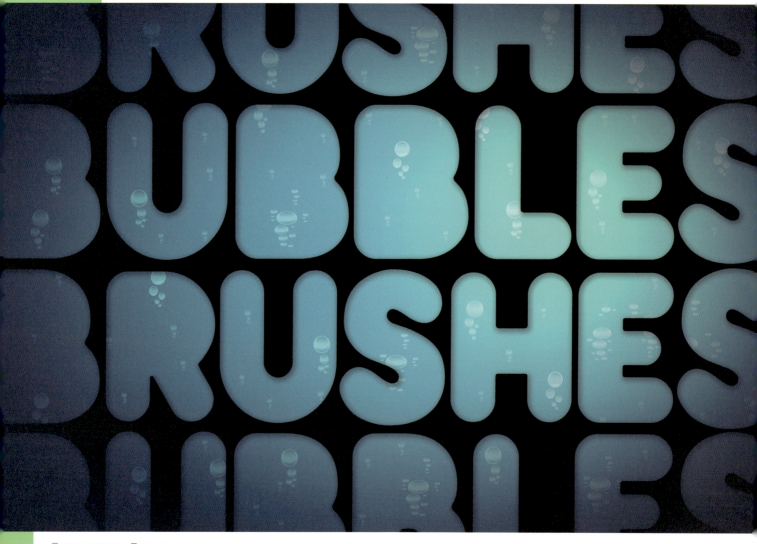

📁 NATURE » 📁 098-Bubbles Brushes

deviantART　http://coby17.deviantart.com/

Alex-Zhang **Number of water**

Brush Number
099

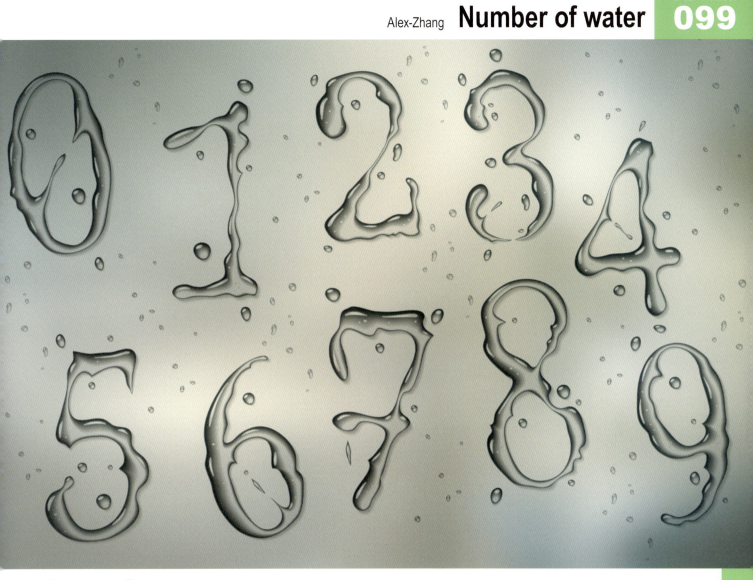

📁 NATURE » 📁 099-Number of water

個人利用 ⭕   商用利用 ⭕ ※3

Brusheezy   http://www.brusheezy.com/members/alex-zhang

NATURE

Brush Number
100 Zummerfish's Ripples Brushes zummerfish

📁 NATURE » 📁 100-Zummerfish's Ripples Brushes

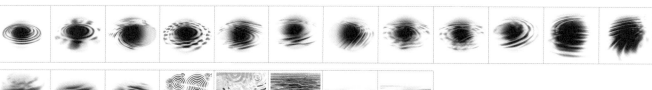

deviantART http://www.zummerfish.deviantart.com/

FackFebruary (Jobey) **Water** — Brush Number **101**

Water splashes

📁 NATURE » 📁 101-Water

個人利用 ○ 商用利用 ○

deviantART http://fackfebruary.deviantart.com/

Brush Number
102 Galaxy Set
Javier Larios Barbosa

NATURE » 102-Galaxy Set

NATURE

deviantART http://javierzhx.deviantart.com/

Tamah! (anodyne-stock) **Starry Night Brush Set**

Brush Number **103**

📁 NATURE 》 📁 103-Starry Night Brush Set

個人利用 ◎   商用利用 ◎

deviantART   http://anodyne-stock.deviantart.com/

Brush Number 104 — StarGlow Brushes _{hawksmont}

Starglow brushes

📁 NATURE » 📁 104-StarGlow Brushes

NATURE

hawksmont Universe :)   http://hawksmont.com/

Brush Number
106 **Stars** hawksmont

📁 NATURE » 📁 106-Stars

個人利用   商用利用

hawksmont Universe :)   http://hawksmont.com/

Coby17 (Brenda Rivera) **Little cute stars Photoshop brushes**

Brush Number 107

NATURE » 107-Little cute stars Photoshop brushes

deviantART http://coby17.deviantart.com/

108 Night Sky Brushes — dauber788

NATURE » 108-Night Sky Brushes

 個人利用 ⃝　 商用利用 ⃝ ※3

Brusheezy　http://www.brusheezy.com/members/dauber788

Macca **1000 Stars** — Brush Number **109**

STAR BRUSH SET

NATURE » 109-1000 Stars

個人利用 ○　商用利用 ○ ※3

Brusheezy　http://www.brusheezy.com/members/macca

NATURE

Brush Number
110 Zummerfish's Nature Brushes zummerfish

📁 NATURE » 📁 110-Zummerfish's Nature Brushes

個人利用 ○   商用利用 ○

deviantART   http://www.zummerfish.deviantart.com

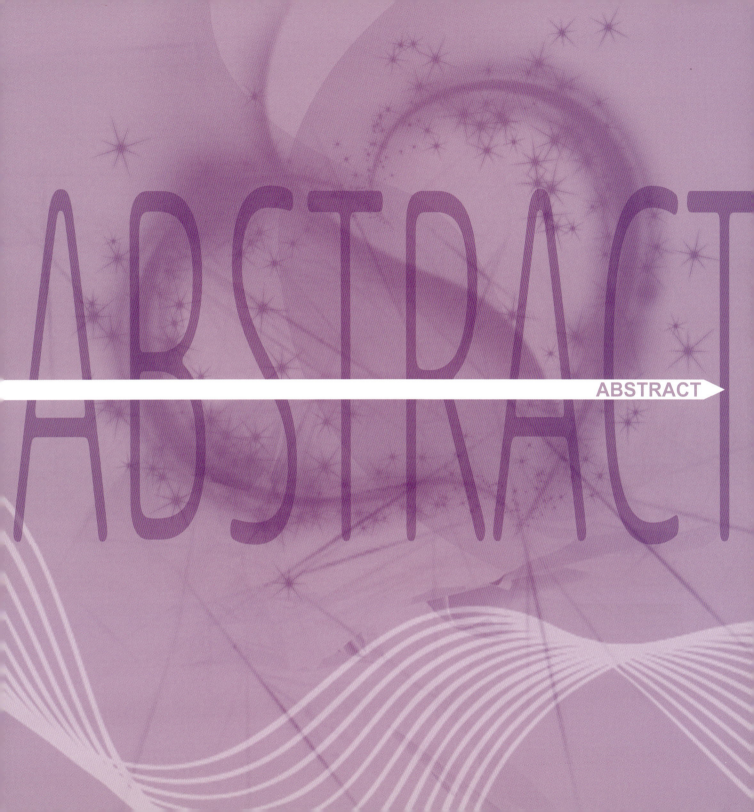

Brush Number
111 horizental-spiral Timothy Blake (Creative Nerds)

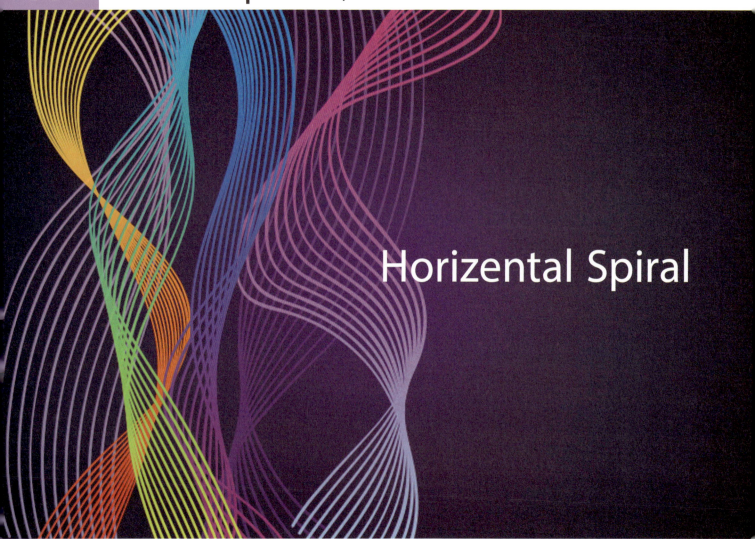

ABSTRACT ≫ 111-horizental-spiral

個人利用 ○ 　　商用利用 △ 有料　　　　　　　　creative NERDS　http://creativenerds.co.uk/

Timothy Blake (Creative Nerds) **Spiral Brush Set**

Brush Number **112**

Looking up!
You'll never find a rainbow if you're looking down.

📁 ABSTRACT ≫ 📁 112-Spiral Brush Set

個人利用 ⃝　　商用利用 ⃝

creative NERDS   http://creativenerds.co.uk/

ABSTRACT

Brush Number
113 Abstract 04 Javier Larios Barbosa

📁 ABSTRACT ≫ 📁 113-Abstract 04

個人利用 ○ 商用利用 △ 有料

deviantART   http://javierzhx.deviantart.com

Javier Larios Barbosa **Abstract 07**

Brush Number **114**

CRYPTID
It has been a strange animal in the unknown world.

📁 ABSTRACT ≫ 📁 114-Abstract 07

個人利用 ○　　商用利用 △　有料

deviantART　http://javierzhx.deviantart.com

ABSTRACT

Brush Number
115 Valentine Glow Hearts hawksmont

📁 ABSTRACT » 📁 115-Valentine Glow Hearts

個人利用 　　商用利用 　　　　　　　Brusheezy　http://www.brusheezy.com/members/hawksmont

ShiftyJ **Gravity Brushes**

Brush Number **116**

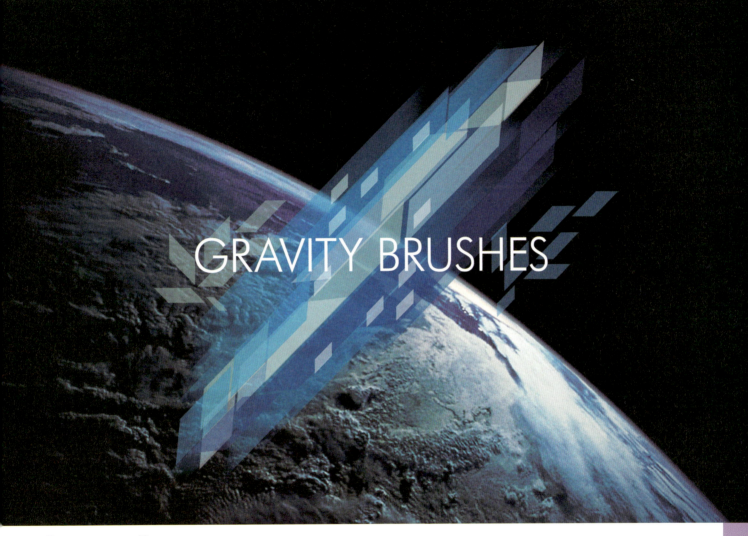

📁 ABSTRACT ≫ 📁 116-Gravity Brushes

deviantART http://shiftyj.deviantart.com/

Brush Number 117 — Kinetic Brush Set *ShiftyJ*

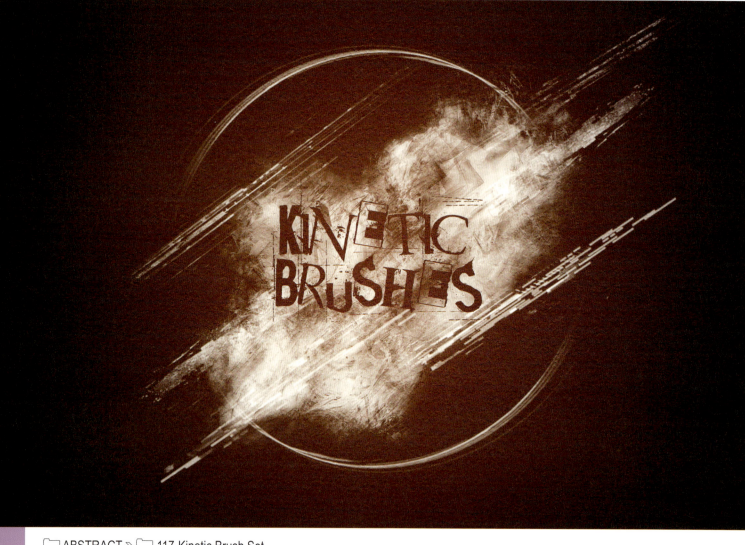

📁 ABSTRACT ≫ 📁 117-Kinetic Brush Set

個人利用 ⭕   商用利用 ⭕ ※2

deviantART   http://shiftyj.deviantart.com/

ShiftyJ Aurora Brushes — Brush Number 118

📁 ABSTRACT ≫ 📁 118-Aurora Brushes

※ 2

deviantART   http://shiftyj.deviantart.com/

Brush Number
119 Fortune Brushes ShiftyJ

📁 ABSTRACT » 📁 119-Fortune Brushes

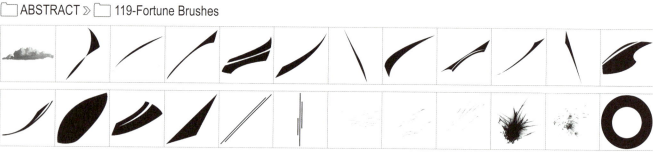

個人利用 ⭕   商用利用 ⭕ ※2

deviantART http://shiftyj.deviantart.com/

ShiftyJ Assassin Brush Set

Brush Number 120

📁 ABSTRACT » 📁 120-Assassin Brush Set

個人利用 ⭕ 商用利用 ⭕ ※2

deviantART   http://shiftyj.deviantart.com/

ABSTRACT

Brush Number
121 Arcade Brushes ShiftyJ

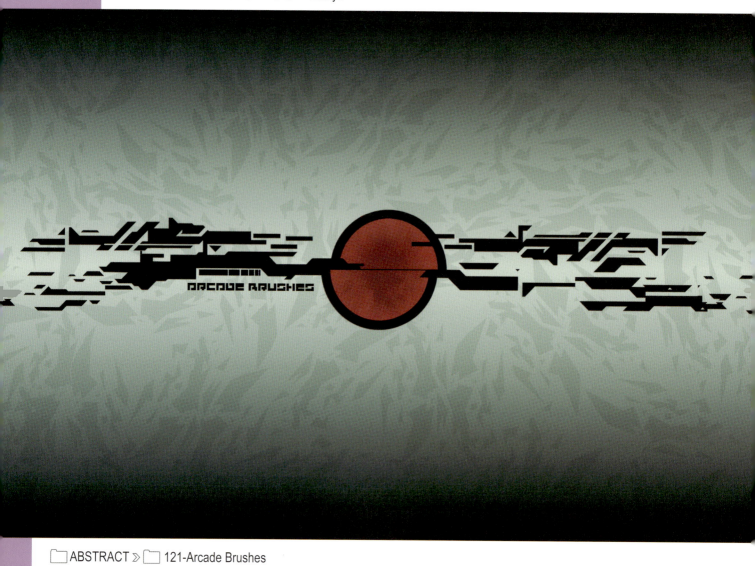

📁 ABSTRACT ≫ 📁 121-Arcade Brushes

 個人利用 ⭕   商用利用 ⭕ ※2

ABSTRACT

deviantART   http://shiftyj.deviantart.com/

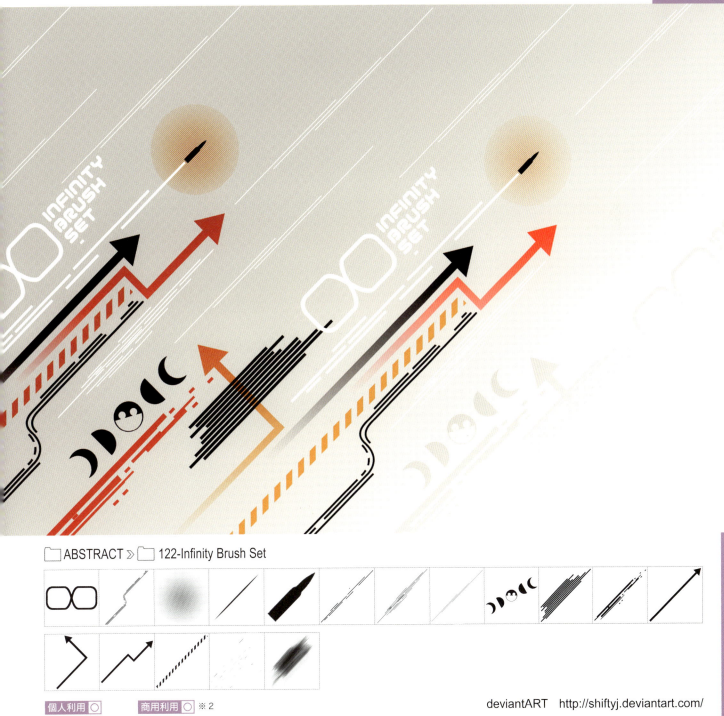

Brush Number
123
extremelyABSTRACT
Przemyslaw 'env1ro' Szczepanski

EXTREMELY ABSTRACT

ABSTRACT ≫ 123-extremelyABSTRACT

ABSTRACT

個人利用 ○   商用利用 ○   *

deviantART   http://env1ro.deviantart.com/

Abstract Paint 2

Przemyslaw 'env1ro' Szczepanski

Brush Number 124

ABSTRACT » 124-Abstract Paint 2

個人利用 ◯　商用利用 ◯ *

deviantART　http://env1ro.deviantart.com/

Brush Number 125 — ABSTRACT FRACTAL
Przemyslaw 'env1ro' Szczepanski

FRACTAL ABSTRACT BY ENV1RO

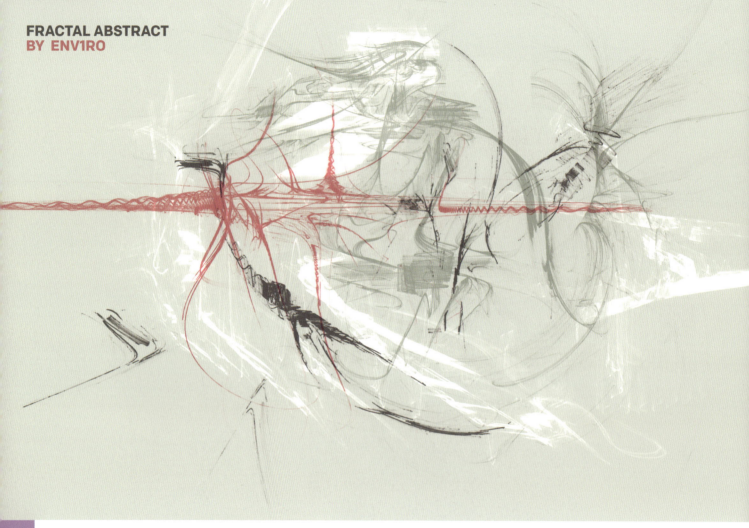

ABSTRACT » 125-ABSTRACT FRACTAL

ABSTRACT

 個人利用 ◯　　 商用利用 ◯ *

deviantART　http://env1ro.deviantart.com/

Coby Sparkles Photoshop Brushes

Coby17 (Brenda Rivera) — Brush Number 126

ABSTRACT » 126-Coby Sparkles Photoshop Brushes

個人利用 ○　　商用利用 ×

deviantART　http://coby17.deviantart.com/

Brush Number
127
15 High Resolution Blingy Photoshop Brushes
Geno Arguelles

BLING BRUSHES

ABSTRACT » 127-15 High Resolution Blingy Photoshop Brushes

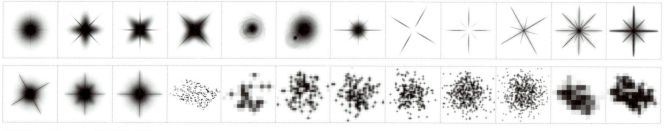

個人利用 ◯　　商用利用 ◯

Geno Arguelles　http://genoarguelles.com/

Timothy Blake (Creative Nerds)　**ink-in-water-photoshop-brush-set**

Brush Number **128**

📁 ABSTRACT ≫ 📁 128-ink-in-water-photoshop-brush-set

個人利用 　　商用利用

The Forgotten Lair　http://creativenerds.co.uk/

ABSTRACT

Brush Number
129 DBD | PatrioticPack Part 1   elsamuel

📁 ABSTRACT » 📁 129-DBD | PatrioticPack Part 1

ABSTRACT

個人利用 ○ ※1   商用利用 ○ ※1   Brusheezy   http://www.brusheezy.com/members/elsamuel

Andrzej Walkowiak **Fireworks high resolution**

Brush Number **130**

FIREWORKS BRUSHS

ABSTRACT » 130-Fireworks high resolution

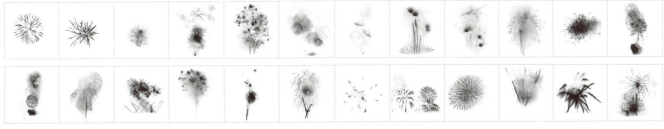

個人利用 ○ 商用利用 ○

deviantART http://lordandre.deviantart.com/

Brush Number
131 Fireworks high resolution 2 Andrzej Walkowiak

FIREWORKS BRUSHS

📁 ABSTRACT » 📁 131-Fireworks high resolution 2

個人利用 ○   商用利用 ○

deviantART   http://lordandre.deviantart.com/

ABSTRACT

Brush Number
132
150 Light Effect Brushes — Jobey Buya

EFFECTS » 132-150 Light Effect Brushes

個人利用 ○   商用利用 ○

Deseign Resources   http://ar-design-resources.blogspot.jp/

Zummerfish's Lightbeams Brushes

zummerfish — Brush Number 133

□ EFFECTS » □ 133-Zummerfish's Lightbeams Brushes

 個人利用   商用利用   deviantART   http://www.zummerfish.deviantart.com/

134 92 Bokeh Brushes — FackFebruary (Jobey)

EFFECTS » 134-92 Bokeh Brushes

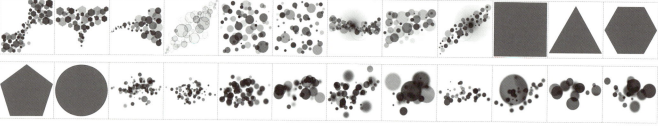

deviantART http://fackfebruary.deviantart.com/

Tamah! (anodyne-stock) **Light Brush Set**

Brush Number **135**

📁 EFFECTS » 📁 135-Light Brush Set

 個人利用 ⭕   商用利用 ⭕

deviantART　http://anodyne-stock.deviantart.com/

Brush Number
136 Lights Effects Brushes PS Coby17 (Brenda Rivera)

📁 EFFECTS » 📁 136-Lights Effects Brushes PS

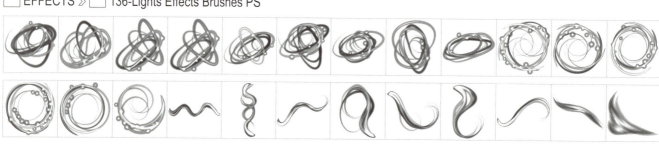

個人利用 ⭕　商用利用 ❌

deviantART　http://coby17.deviantart.com/

SparkleStock **12 Large Bokeh Brushes**

Brush Number
137

☐ EFFECTS » ☐ 137-12 Large Bokeh Brushes

SparkleStock http://www.sparklestock.com/

EFFECTS

138 — 16 Photorealistic Explosion Brushes — SparkleStock

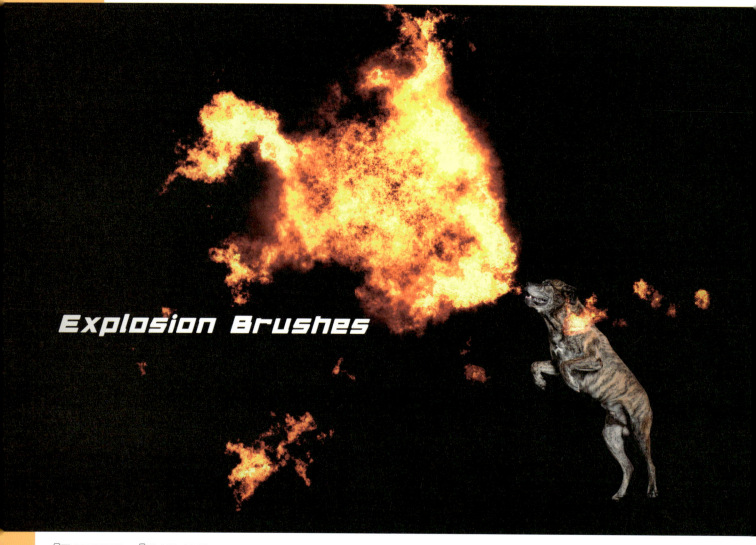

📁 EFFECTS » 📁 138-16 Photorealistic Explosion Brushes

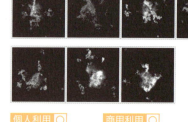

個人利用 ◯　商用利用 ◯

SparkleStock　http://www.sparklestock.com/

musicfreak469596 **Bokeh Heart Brush Pack**

Brush Number **139**

📁 EFFECTS » 📁 139-Bokeh Heart Brush Pack

個人利用 ○ ※1　商用利用 ○ ※1

Brusheezy　http://www.brusheezy.com/members/musicfreak469596

EFFECTS

Brush Number
140
Brushpack Jon Bee

Lighting PRO
Our indirect lighting fixtures are pleasing

📁 EFFECTS » 📁 140-Brushpack

EFFECTS

Simen91 **Simen 91's Star and Light-effect Brushes**

Brush Number **141**

All journeys have secret destinations of which the traveler is unaware **Starlight Ship**

📁 EFFECTS » 📁 141-Simen 91's Star and Light-effect Brushes

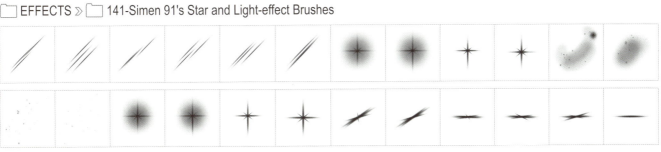

個人利用 ⭕ ※2　商用利用 ❌

Brusheezy　http://www.brusheezy.com/members/simen91

EFFECTS

Brush Number
142 Circles Madness brushes
Przemyslaw 'env1ro' Szczepanski

📁 EFFECTS ≫ 📁 142-Circles Madness brushes

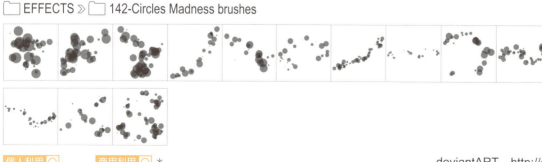

EFFECTS

個人利用 ⭕   商用利用 ⭕ *

deviantART http://env1ro.deviantart.com/

Brush Number
143 DIRT AND GRIME PHOTOSHOP BRUSH SET
Nathan Brown

GRUNGE » 143-DIRT AND GRIME PHOTOSHOP BRUSH SET

GraphicMonkee   http://www.graphicmonkee.com/

Timothy Blake (Creative Nerds) **concreate-brush-set**

Brush Number **144**

📁 GRUNGE » 📁 144-concreate-brush-set

個人利用 ☐ 商用利用 ☐

creative NERDS　http://creativenerds.co.uk/

GRUNGE

Brush Number 145 — Demolished Cracks
Przemyslaw 'env1ro' Szczepanski

📁 GRUNGE » 📁 145-Demolished Cracks

deviantART　http://env1ro.deviantart.com/

Timothy Blake (Creative Nerds) **grundges-frames-brushes**

Brush Number **146**

📁 GRUNGE » 📁 146-grundges-frames-brushes

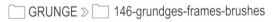

GRUNGE

creative NERDS　　http://creativenerds.co.uk/

Brush Number **147** Twisted Mind Cracked brushes Vol 1 Michael Kelley

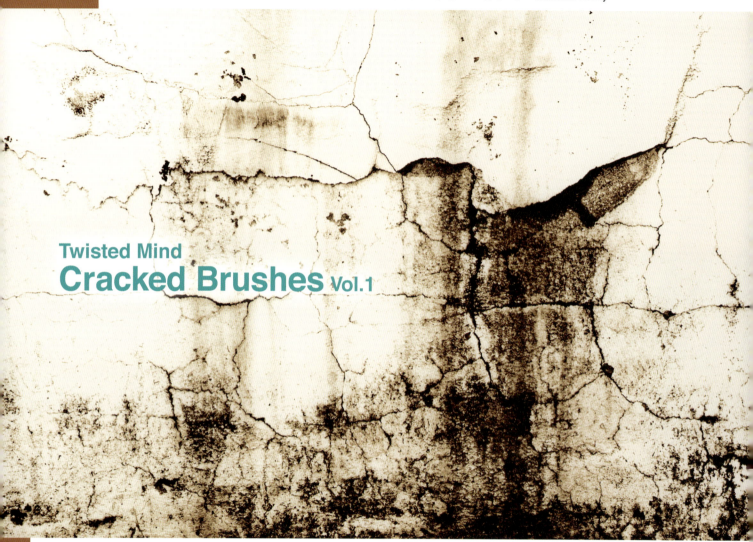

📁 GRUNGE ≫ 📁 147-Twisted Mind Cracked brushes Vol 1

GRUNGE

deviantART http://textures-and-more.deviantart.com/

Twisted Mind Cracked brushes Vol 2

Michael Kelley

Brush Number **148**

GRUNGE » 148-Twisted Mind Cracked brushes Vol 2

deviantART http://textures-and-more.deviantart.com/

Brush Number
149 Twisted Mind peeling paint vol 1 Michael Kelley

GRUNGE » 149-Twisted Mind peeling paint vol 1

GRUNGE

 個人利用 ◯   商用利用 ◯

deviantART http://textures-and-more.deviantart.com/

hawksmont Crack Brushes I

Brush Number 150

📁 GRUNGE ≫ 📁 150-Crack Brushes I

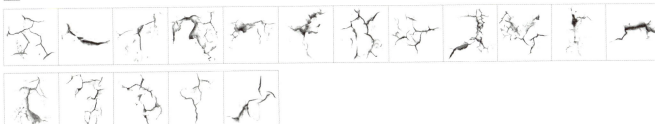

deviantART　http://hawksmont.deviantart.com/

GRUNGE

Brush Number
151 Cracks Part III hawksmont

GRUNGE » 151-Cracks Part III

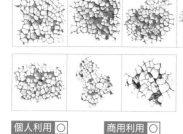

個人利用 ○   商用利用 ○

deviantART   http://hawksmont.deviantart.com/

Tijo **42 Grubby Grunge Brushes**

Brush Number **152**

📁 GRUNGE » 📁 152-42 Grubby Grunge Brushes

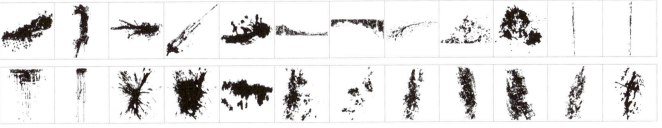

個人利用 　商用利用

Brusheezy　http://www.brusheezy.com/members/tijo

GRUNGE

Brush Number
153 Adobe Photoshop Grunge Brushes 2012 James Myers (nineteeneightysevendesign)

📁 GRUNGE » 📁 153-Adobe Photoshop Grunge Brushes 2012

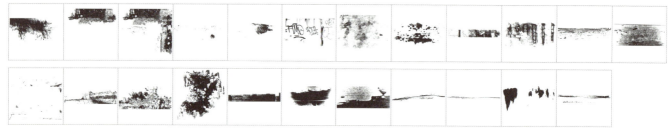

GRUNGE

個人利用 ◯ ※2   商用利用 ✕   Brusheezy   http://www.brusheezy.com/members/nineteeneightysevendesign

James Myers（nineteeneightysevendesign） **Grunge brush set (image pack included!)**

Brush Number **154**

📁 GRUNGE ≫ 📁 154-Grunge brush set (image pack included!)

個人利用 ◯ ※2　商用利用 ✕　　　Brusheezy　http://www.brusheezy.com/members/nineteeneightysevendesign

GRUNGE

Brush Number
155 Grunge Set 4 James Myers (nineteeneightysevendesign)

📁 GRUNGE » 📁 155-Grunge Set 4

Brusheezy http://www.brusheezy.com/members/nineteeneightysevendesign

Brush Number
156
Vector 03 Javier Larios Barbosa

deviantART http://javierzhx.deviantart.com/

Javier Larios Barbosa **Vector 05**

Brush Number **157**

VECTOR 05

📁 VECTOR » 📁 157-Vector 05

deviantART http://javierzhx.deviantart.com/

VECTOR

Brush Number
158 Hearts hawksmont

📁 VECTOR » 📁 158-Hearts

deviantART http://hawksmont.deviantart.com/

hawksmont Ribbon Bow Brushes

Brush Number 159

VECTOR » 159-Ribbon Bow Brushes

個人利用 ◯ 商用利用 ◯

deviantART http://hawksmont.deviantart.com/

VECTOR

Brush Number
160
Wireframe Brushes Javier Larios Barbosa

📁 VECTOR » 📁 160-Wireframe Brushes

VECTOR

個人利用   商用利用

deviantART http://javierzhx.deviantart.com/

Coby17 (Brenda Rivera) **Bird Cage Photoshop Brushes** Brush Number **161**

📁 VECTOR » 📁 161-Bird Cage Photoshop Brushes

deviantART http://coby17.deviantart.com/

Brush Number
162 Shiny Circles Brushes Coby17 (Brenda Rivera)

📁 VECTOR » 📁 162-Shiny Circles Brushes

個人利用 ○   商用利用 ×

deviantART   http://coby17.deviantart.com/

elsamuel **DBD | VectorPack Brushes**

Brush Number **163**

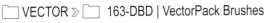

VECTOR » 163-DBD | VectorPack Brushes

個人利用 ◯ 商用利用 ◯ ※1

Brusheezy http://www.brusheezy.com/members/elsamuel

VECTOR

Brush Number
164 Ribbons 3 titimontoya

VECTOR » 164-Ribbons 3

VECTOR

個人利用 ○ 商用利用 ○ ※1

Brusheezy   http://www.brusheezy.com/members/titimontoya

Ribbons 4

titimontoya

Brush Number 165

📁 VECTOR » 📁 165-Ribbons 4

個人利用 ⭕   商用利用 ⭕ ※1

Brusheezy　http://www.brusheezy.com/members/titimontoya

VECTOR

Brush Number
166 Variety Brushies1 titimontoya

📁 VECTOR » 📁 166-Variety Brushies1

VECTOR

Brusheezy　http://www.brusheezy.com/members/titimontoya

titimontoya **Geo5**

Brush Number **167**

📁 VECTOR » 📁 167-Geo5

個人利用 ⭕ 商用利用 ⭕ ※1

Brusheezy   http://www.brusheezy.com/members/titimontoya

VECTOR

Brush Number
168 Line Art Abstract Brushies (Line Art1,2,3) titimontoya

LINE ART ABSTRACT BRUSHES

VECTOR » 168-Line Art Abstract Brushies

VECTOR

Brusheezy　http://www.brusheezy.com/members/titimontoya

Brush Number 169 — Lace and Floral dividers Photoshop brushes — Coby17 (Brenda Rivera)

ORNAMENT » 169-Lace and Floral dividers Photoshop brushes

deviantART http://coby17.deviantart.com/

Coby17 (Brenda Rivera) **Circular Laces Photoshop Brushes**

Brush Number
170

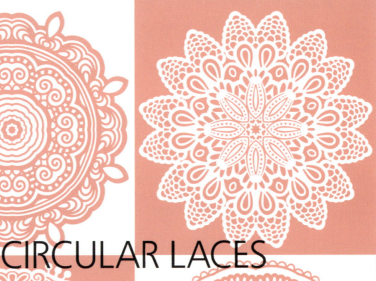

CIRCULAR LACES

📁 ORNAMENT » 📁 170-Circular Laces Photoshop Brushes

deviantART   http://coby17.deviantart.com/

Brush Number 171: Laces Brushes — Coby17 (Brenda Rivera)

ORNAMENT » 171-Laces Brushes

deviantART http://coby17.deviantart.com/

Coby17 (Brenda Rivera) **Circular laces HQ Photoshop Brushes**

Brush Number
172

📁 ORNAMENT ≫ 📁 172-Circular laces HQ Photoshop Brushes

個人利用 ⭕   商用利用 ❌

deviantART   http://coby17.deviantart.com/

Brush Number 173

Fansy Lace Brushes
Coby17 (Brenda Rivera)

ORNAMENT » 173-Fansy Lace Brushes

deviantART http://coby17.deviantart.com/

Coby17 (Brenda Rivera) **Kawaii Pixel Laces Brushes**

Brush Number **174**

📁 ORNAMENT » 📁 174-Kawaii Pixel Laces Brushes

deviantART http://coby17.deviantart.com/

Brush Number
175
Vintage Ornaments and Lace Photoshop Brushes Coby17 (Brenda Rivera)

📁 ORNAMENT » 📁 175-Vintage Ornaments and Lace Photoshop Brushes

ORNAMENT

個人利用 　商用利用

deviantART　http://coby17.deviantart.com/

Jan Willem Geertsma · **Ornament Brushes 1** · Brush Number **176**

📁 ORNAMENT » 📁 176-Ornament Brushes 1

個人利用 ○　商用利用 ○

GEERTSMA.NL　http://www.geertsma.nl/

Brush Number 177 — Ornament Brushes 2 Jan Willem Geertsma

ORNAMENT » 177-Ornament Brushes 2

個人利用 ○ 商用利用 ○ GEERTSMA.NL http://www.geertsma.nl/

Alex-Zhang **Decorative patterns**

Brush Number **178**

 ORNAMENT » 178-Decorative patterns

Brusheezy   http://www.brusheezy.com/members/alex-zhang

Brush Number
179 Mix 13 — Mohaafterdark

📁 ORNAMENT ≫ 📁 179-Mix 13

個人利用 ○　※　商用利用 ×

POLICE Media　http://mohaafterdark.blogspot.jp/

Mohaafterdark **MIX 7** Brush Number **180**

📁 ORNAMENT » 📁 180-MIX 7

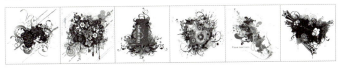

 ※

POLICE Media http://mohaafterdark.blogspot.jp/

Brush Number
181 **MIX 3** Mohaafterdark

📁 ORNAMENT » 📁 181-MIX 3

ORNAMENT

個人利用 ※   商用利用

POLICE Media   http://mohaafterdark.blogspot.jp/

Mohaafterdark **MIX 4** — Brush Number **182**

Mix 4 brushes

📁 ORNAMENT » 📁 182-MIX 4

個人利用 ◯ ※   商用利用 ✕

POLICE Media http://mohaafterdark.blogspot.jp/

ORNAMENT

Brush Number
183 patterns borders Mohaafterdark

ORNAMENT » 183-patterns borders

ORNAMENT

個人利用 ○ ※   商用利用 ✕

POLICE Media   http://mohaafterdark.blogspot.jp/

Mohaafterdark **Ornate** — Brush Number **184**

ORNATE BRUSHES

ORNAMENT » 184-Ornate

個人利用 ○ ※　商用利用 ✕

POLICE Media　http://mohaafterdark.blogspot.jp/

Brush Number 185

Floral 2 Mohaafterdark

Lorem ipsum dolor sit amet, consectetur adipiscing elit. Vivamus dui lorem, hendrerit eu rutrum ut, dapibus dignissim ipsum. Nam ut leo rhoncus, scelerisque purus ut, lobortis turpis. Ut dapibus, nunc ut ornare facilisis, purus nisl efficitur tortor, nec ullamcorper elit eros sed risus. Duis quis purus lorem. Praesent et nisl metus. Donec lectus magna, mattis in nunc id, vehicula porta erat. Curabitur in tempus mi. In elementum augue nec ante tincidunt venenatis. Etiam egestas dolor lorem.

Lorem ipsum dolor sit amet, consectetur adipiscing elit. Vivamus dui lorem, hendrerit eu rutrum ut, dapibus dignissim ipsum. Nam ut leo rhoncus, scelerisque purus ut, lobortis turpis. Ut dapibus, nunc ut ornare facilisis, purus nisl efficitur tortor, nec ullamcorper elit eros sed risus. Duis quis purus lorem. Praesent et nisl metus. Donec lectus magna, mattis in nunc id, vehicula porta erat. Curabitur in tempus mi. In elementum augue nec ante tincidunt venenatis. Etiam egestas dolor lorem.

個人利用 ◯ ※   商用利用 ✕

POLICE Media   http://mohaafterdark.blogspot.jp/

Brush Number
186 Branches hawksmont

VARIETY » 186-Branches

deviantART http://hawksmont.deviantart.com/

hawksmont **Cherry Blossom 1**

Brush Number **187**

Cherry blossom
Brushes

📁 VARIETY » 📁 187-Cherry Blossom 1

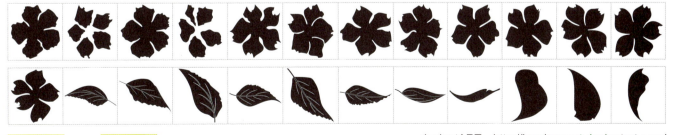

個人利用 ◯   商用利用 ◯

deviantART   http://hawksmont.deviantart.com/

Brush Number
188 Water Pack 2 — Delia Galhotra

Water Pack Photoshop Brushes

 VARIETY » 188-Water Pack 2

VARIETY

 個人利用 ◯   商用利用 ✕   Delia Galhotra freelance phtographer   http://deliagalhotra.wix.com/photography/

hawksmont **Spiderz Brushes**

Brush Number **189**

VARIETY » 189-Spiderz Brushes

deviantART http://hawksmont.deviantart.com/

Brush Number
190 SpiderWebBrushes Cary_HMS

📁 VARIETY » 📁 190-SpiderWebBrushes

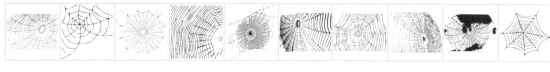

Brusheezy   http://www.brusheezy.com/members/cary_hms

hawksmont **Butterfly Brushes I**

Brush Number **191**

📁 VARIETY » 📁 191-Butterfly Brushes I

deviantART   http://hawksmont.deviantart.com/

Brush Number
192 Butterfly Brushes II hawksmont

VARIETY » 192-Butterfly Brushes II

deviantART http://hawksmont.deviantart.com/

Brush Number
194 Skylines hawksmont

📁 VARIETY » 📁 194-Skylines

deviantART http://hawksmont.deviantart.com/

Coby17 (Brenda Rivera) **Hand draws borders brushes for Photoshop**

Brush Number **195**

Hand Draws Borders

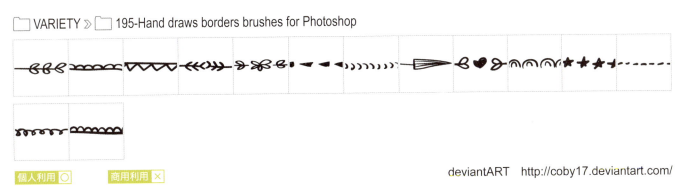

VARIETY ≫ 195-Hand draws borders brushes for Photoshop

個人利用 ◯   商用利用 ✕

deviantART   http://coby17.deviantart.com/

Digital Brushs — Coby17 (Brenda Rivera)

Brush Number 196

VARIETY » 196-Digital Brushs

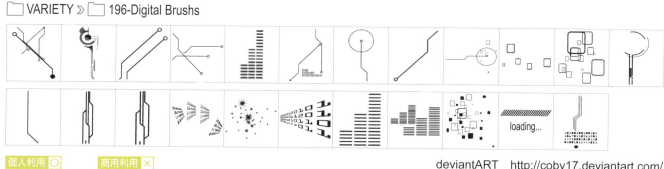

deviantART http://coby17.deviantart.com/

Alex-Zhang **Technology**

Brush Number
197

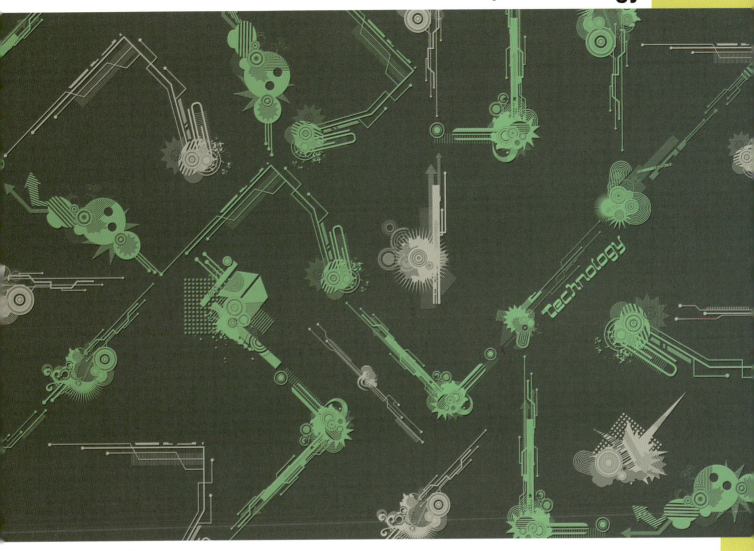

VARIETY » 197-Technology

個人利用 商用利用 ※3

Brusheezy http://www.brusheezy.com/members/alex-zhang

VARIETY

Brush Number
198
23 Free HUD Photoshop Brushes — Nino Batitis

📁 VARIETY » 📁 198-23 Free HUD Photoshop Brushes

  Mac Only

YouTheDesigner http://www.youthedesigner.com/

Lego Brick Brush Pack

eflouret

Brush Number 199

📁 VARIETY » 📁 199-Lego Brick Brush Pack

個人利用 ◯ 商用利用 ◯ ※1

Brusheezy　http://www.brusheezy.com/members/eflouret

Brush Number **200**

50 Puzzle Pieces Brushes Cary_HMS

VARIETY » 200-50 Puzzle Pieces Brushes

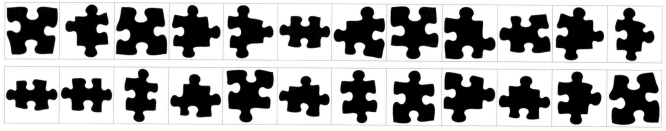

 ※3

Brusheezy   http://www.brusheezy.com/members/cary_hms

Cary_HMS **11Cool Angle Wheels**

Brush Number **201**

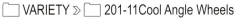 VARIETY » 201-11Cool Angle Wheels

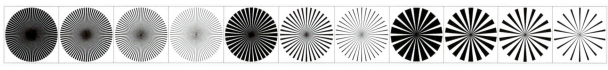

  ※3

Brusheezy   http://www.brusheezy.com/members/cary_hms

Brush Number
202
Rain - 4 Brushes Jon Bee

📁 VARIETY » 📁 202-Rain - 4 Brushes

VARIETY

個人利用 ◯   商用利用 ◯

Przemyslaw 'env1ro' Szczepanski **FunkySUNBURST**

Brush Number **203**

📁 VARIETY » 📁 203-FunkySUNBURST

個人利用 ◯　商用利用 ◯　*

deviantART　http://env1ro.deviantart.com/

VARIETY

Brush Number
204
Skin Brushes
Przemyslaw 'env1ro' Szczepanski

VARIETY » 204-Skin Brushes

deviantART http://env1ro.deviantart.com/

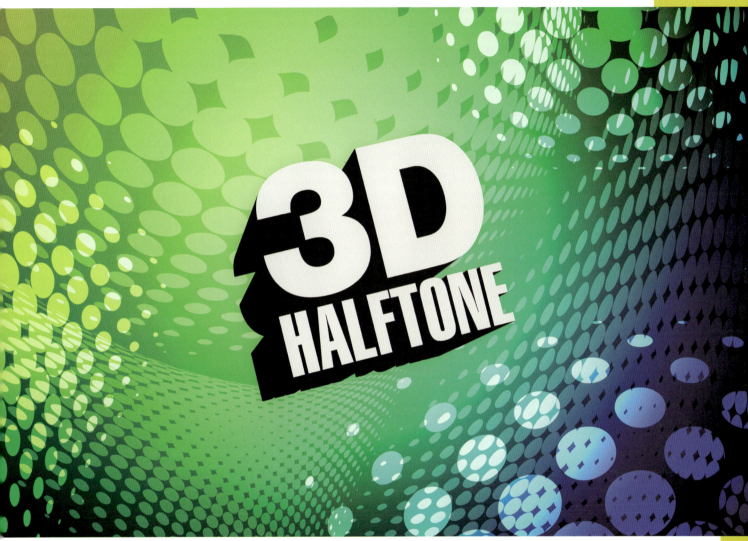

Przemyslaw 'env1ro' Szczepanski — **3D Halftone brushes** — Brush Number **205**

VARIETY » 205-3D Halftone brushes

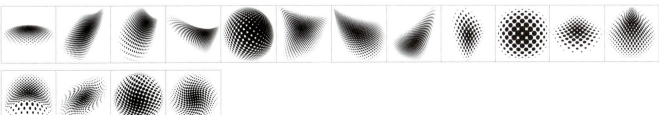

deviantART http://env1ro.deviantart.com/

Brush Number
206
HALFTONEThree
Przemyslaw 'env1ro' Szczepanski

📁 VARIETY » 📁 206-HALFTONEThree

個人利用 ◯　商用利用 ◯ *

deviantART　http://env1ro.deviantart.com/

Halftone 2 brushes

Brush Number **207**

Przemyslaw 'env1ro' Szczepanski

📁 VARIETY ≫ 📁 207-Halftone 2 brushes

  *

deviantART   http://env1ro.deviantart.com/

Brush Number
208
CoffeeHappens RELOADED
Przemyslaw 'env1ro' Szczepanski

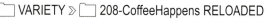

VARIETY » 208-CoffeeHappens RELOADED

  *

deviantART   http://env1ro.deviantart.com/

HighTech Circles

Przemyslaw 'env1ro' Szczepanski

Brush Number **209**

VARIETY ≫ 209-HighTech Circles

  *

deviantART　http://env1ro.deviantart.com/

Brush Number
210
RisingSun brushes
Przemyslaw 'env1ro' Szczepanski

📁 VARIETY » 📁 210-RisingSun brushes

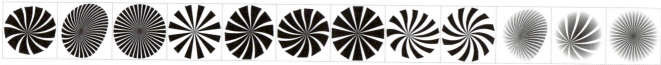

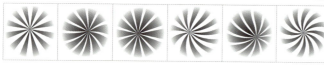

  *

deviantART http://env1ro.deviantart.com/

natieditions00 **12 Feathers Brushes**

Brush Number **211**

VARIETY » 211-12 Feathers Brushes

deviantART http://natieditions00.deviantart.com

Brush Number
212
10 Trail To Nowhere   Phoenix

10 tail to nowhere brushes

📁 VARIETY » 📁 212-10 Trail To Nowhere

個人利用 ○   商用利用 ○

The Forgotten Lair   http://www.theforgottenlair.net/

FackFebruary (Jobey) **42 Shattered Glass Brushes**

Brush Number **213**

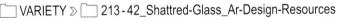

VARIETY » 213 - 42_Shattred-Glass_Ar-Design-Resources

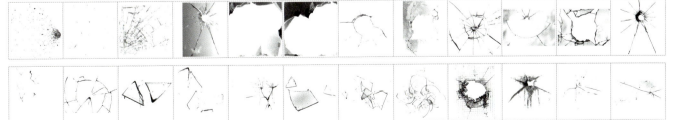

 個人利用
 商用利用

deviantART http://fackfebruary.deviantart.com/

Brush Number
214 Broken Glass Brushes - Eight Jon Bee

VARIETY » 214-Broken Glass Brushes - Eight

FackFebruary (Jobey) **118 Fog Brushes** — Brush Number 215

VARIETY » 215-118 Fog Brushes

deviantART http://fackfebruary.deviantart.com/

デザインに即戦力
Photoshop 厳選ブラシ集

2015 年 3 月 31 日　初版第 1 刷発行

制作　ラトルズ編集部

発行者　黒田庸夫
発行所　株式会社ラトルズ
〒 102-0083　東京都千代田区麹町 1-8-14 麹町 YK ビル 3 階
TEL　03-3511-2785　　FAX　03-3511-2786
http://www.rutles.net

印刷・製本　株式会社ルナテック

ISBN978-4-89977-432-7
Copyright ©2015
Printed in Japan

【お断り】
- 本書の一部または全部を無断で複写複製することは、法律で認められた場合を除き、著作権の侵害となります。
- 本書に関してご不明な点は、当社 Web サイトの「ご質問・ご意見」ページ (http://www.rutles.net/contact/index.php) をご利用ください。
　電話、ファックス、電子メールでのお問い合わせには応じておりません。
- 当社への一般的なお問い合わせ は、info@rutles.net または上記の電話、ファックス番号までお願いいたします。
- 本書内容については、間違いがないよう最善の努力を払って検証していますが、著者および発行者は、本書の利用によって生じたいかなる障害に対してもその責を負いませんので、あらかじめご了承ください。
- 乱丁、落丁の本が万一ありましたら、小社営業宛てにお送りください。送料小社負担にてお取り替えします。